SECOND EDITION

Safety First Checklist

Audit & Inspection Program for Children's Play Areas

Sally McIntyre and Susan M. Goltsman

MIG Communications
Berkeley, California

MIG Communications, 800 Hearst Avenue, Berkeley, CA 94710, USA
(510) 845-0953; fax (510) 845-8750

© 1989, 1997 by MIG Communications. All rights reserved.
Published 1989, Second Edition 1997.
Printed in the United States of America.

No part of this work covered by the copyright hereon may be reproduced or used in any form or by any means—graphic, electronic, or mechanical, including photocopying, recording, taping, or information storage and retrieval systems—without written permission of the publisher.

Editors: Paul Yee, David Driskell
Cover Design: MIG DesignWorks

Library of Congress Cataloging-in-Publication Data
McIntyre, Sally.
　　Safety first checklist : audit and inspection program for
　　children's play areas / Sally McIntyre, Susan M. Goltsman -- 2nd
　　ed.
　　　　p.　　cm.
　　Includes bibliographical references.
　　094466119X (pbk.)
　　　1. Playgrounds--Equipment and supplies--Safety measures.
　　2. Playgrounds--Safety measures--Evaluation.　　I. Goltsman, Susan M.
　　II. Title.
　　GV426.5.M4　1997　　　　　　　　　　　　　　　　95-18726
　　796'.06'8028--DC20　　　　　　　　　　　　　　　　CIP

Illustration Credits
The following illustrations were adapted with permission, and should not be reprinted without permission of the copyright holder:

Copyright, American Society for Testing and Materials: Figures 1, 2, 4–8, 10, 13–22, 25–28.

Copyright, Landscape Structures, Inc.: the illustrations in Checklists 3, 5–10, 12, 13, 15–21.

Copyright, Kompan, Inc.: the illustrations in Checklists 11, 14.

CONTENTS

To Safety First Users	iv
About the Authors	v
Acknowledgements	vi
About MIG	vi
Introduction	1
Definitions	7
Inspection Tools	14
Inspection Procedures	16

Safety Checklists
  Daily Inspection Checklist
  1   Site Survey
  2   Safety Surfacing
  3   Equipment Access & Egress
  4   Guardrails & Protective Barriers
  5   Balance Beams
  6   Bars, Chin-up & Turning
  7   Bars, Parallel
  8   Bridges, Clatter
  9   Bridges, Stationary
  10   Climbers
  11   Climbers, Flexible
  12   Fire Poles
  13   Horizontal Ladders & Ring Treks
  14   Playhouses
  15   Slides
  16   Spring Rocking Equipment
  17   Swings
  18   Swings, Rotating
  19   Track Rides
  20   Tunnels
  21   Composite Structures

Safety Inspection Summary	121
Bibliography	122

LONGWOOD COLLEGE LIBRARY
FARMVILLE, VIRGINIA 23901

TO SAFETY FIRST USERS

USE AGREEMENT FOR SAFETY FIRST CHECKLIST

Copyright. *Safety First Checklist* is the copyrighted property of MIG Communications and is protected by United States copyright laws and international treaty provisions.

Grant of License. MIG Communications grants to you, the end user (either an individual or entity), the right to use *Safety First Checklist* forms included herein to conduct a safety evaluation of a single outdoor facility. In granting this right, MIG Communications gives permission for *Safety First Checklist* forms to be reproduced only to the extent necessary for complete evaluation of the aforementioned single outdoor facility.

Other Restrictions. This Use Agreement is your proof of license to exercise the rights granted herein and must be retained by you.

Should you have any questions concerning this agreement, or if you desire to contact us for any reason, please write MIG Communications, 800 Hearst Avenue, Berkeley, CA 94710, or call (510) 845-0953.

MIG Communications is committed to providing you with up-to-date information in a format that is understandable and usable. The information contained in this book is based on ASTM standards (F 1487, F 1292) and U.S. Consumer Product Safety Commission Guidelines (Pub. 325), and was current as of March 1997. It is your responsibility to review the information contained in *Safety First Checklist* in light of the most recent codes and standards applicable to your jurisdiction. In addition, please note the following:

Limits of Liability. The authors and publishers of this document assume no risk or liability for incidents arising from the application of this information in any way. The document should not be construed as a substitute for ASTM standards (F 1487, F 1292), U.S. Consumer Product Safety Commission Guidelines (Pub. 325), or the applicable play area safety codes or standards of the state or jurisdiction in which the document is used.

Staff Training. The National Recreation and Park Association (NRPA) offers a National Playground Safety Inspector Course in locations across the country. Advanced reading and ten hours of training from nationally known playground safety experts will prepare you to take the Certified Playground Safety Inspector Examination offered at the culmination of the certification course. Call (703) 820-4940 for more information.

Evaluating *Safety First*. MIG Communications will be updating this edition periodically to reflect new research and to better meet the needs of safety and maintenance staff. Please send your comments and ideas for revisions, as well as user experience, to MIG Communications, 800 Hearst Avenue, Berkeley, CA 94710.

ABOUT THE AUTHORS

SALLY MCINTYRE

Sally McIntyre, CLP, RTR, CPSI, is the principal of the Eugene, Oregon, office of Moore Iacofano Goltsman, Inc. She is a recreational therapist and recreational planner with more than fourteen years experience as a consultant on a broad range of policy, program, operations, and design issues that impact community livability. She has trained hundreds of private-sector, school district, municipal, and federal employees in the United States as well as abroad in issues relating to the safety and maintenance of recreational environments.

An accomplished technical writer and editor, Ms. McIntyre is the author of two technical manuals produced for the U.S. armed forces to improve play area safety nationwide. She is certified by the National Recreation and Park Association as a national playground safety inspector and is a trainer for the NRPA National Playground Safety Institute. Ms. McIntyre currently serves on the board of directors of the Oregon Recreation and Park Association, and is a member of the National Recreation and Park Association's Pacific Northwest Regional Council.

SUSAN M. GOLTSMAN

Susan Goltsman, ASLA, is a partner with Moore Iacofano Goltsman, Inc., and a cofounder and director of PLAE, Inc. An internationally recognized expert in children's environmental design and planning, she holds degrees in landscape architecture, interior design, and environmental psychology. For the past sixteen years, she has been involved in creating policy, programs, and special environments that promote the development of children, youth, and families. Ms. Goltsman was appointed to the U.S. Architectural and Transportation Barriers Compliance Board (ATBCB) to serve on the committee that is developing ADA guidelines for recreational and outdoor environments. She is a past board-member of the American Chapter of the International Association for the Child's Right to Play and the Center for Childhood.

Ms. Goltsman has written many articles promoting design that responds to the needs of children and youth, and has lectured extensively throughout the country. She is coauthor of the *Play For All Guidelines,* which has been internationally adopted as a reference on the planning, design, and management of outdoor play settings for all children. Ms. Goltsman also coauthored *The Accessibility Checklist,* an evaluation system for buildings and outdoor settings. She is an adjunct faculty member for the Program on Urban Studies at Stanford University.

ACKNOWLEDGEMENTS

All work builds on past contributions to the field. In updating the *Safety First Checklist,* we would like to acknowledge the contributions made to the field of playground safety by the following individuals: Steve King, Dr. Fran Wallach, John Preston, Teri Hendy, Ken Kutska, Kevin Hoffman, Jay Beckwith, and Dr. Seymour Gold. Thank you also to national organizations that have improved the lives of our children through their work in promoting playground safety: the American Society for Testing and Materials, the National Recreation and Park Association, and the National Safety Council. Thank you to Ed Racht, Bob Riffel, and the U.S. Army Corps of Engineers for their support of research that has advanced the field and will improve the safety of children nationwide. A special thank you to Daniel Iacofano for his continuing support of our work, and to the community environmental design division at MIG, Inc., our friends and colleagues. For their assistance in preparing this edition, thank you to Landscape Structures, Inc., and Kompan, Inc., for permission to adapt illustrations for use in this manual; and to the American Society for Testing and Materials, for permission to adapt illustrations and to incorporate ASTM standards into the checklist questions presented in this manual.

Finally, we would like to dedicate this book to all the hundreds of children with and without disabilities in Project PLAE at the Washington Environmental Yard in Berkeley—including Charlie, Emily, Jean, Shoshanna, and Dolly. These children and their families taught us what we really needed to learn.

Susan Goltsman and Sally McIntyre

ABOUT MIG

Moore Iacofano Goltsman (MIG), Inc., is a firm of recreation specialists, landscape architects, planners, social scientists, designers, and communications experts. We offer a full range of services, including: universal design; public involvement and information; park and recreation master planning; ADA transition plans; children's environments; recreation programming; staff training; plan checking for safety and accessibility; and benefits-based planning and design.

In all our work, MIG is committed to planning and design that supports human development for people of all ages and abilities. MIG has offices in Berkeley, California; Los Angeles; Eugene, Oregon; and Raleigh, North Carolina.

INTRODUCTION

WHAT IS THE SAFETY FIRST CHECKLIST?

The *Safety First Checklist* translates the most up-to-date information on playground safety into an easy-to-use play area inspection system. The *Checklist* is based on ASTM F 1487-95 and the U.S. Consumer Product Safety Commission's *Handbook for Public Playground Safety* (CPSC, 1994). The American Society for Testing and Materials' (ASTM) standard is a voluntary standard intended primarily as a guide for equipment manufacturers. The CPSC guidelines are intended as a guide for professionals and members of the general public who are concerned with playground safety. Both documents address "public" playground equipment installed in parks, schools, childcare facilities, institutions, multiple-family dwellings, restaurants, recreational developments, and other public areas. The *Safety First Checklist* does not apply to amusement park equipment, exercise equipment, home playground equipment, or soft-contained play equipment. In some instances where guidance is not provided by CPSC or ASTM F 1487, the authors provide additional guidance. Safety criteria not addressed by CPSC or ASTM F 1487 are identified as the authors' opinion.

The *Safety First Checklist* includes:

- definitions of safety terms;
- a list of tools needed to conduct an inspection;
- a description of inspection procedures;
- a daily inspection checklist;
- a general site survey;
- a surfacing evaluation;
- modular inspection forms for each type of play equipment (swings, slides, climbers, etc.);
- a sample inspection summary form to document inspection results; and
- a bibliography.

The Daily Inspection Checklist and the provision of separate audit, periodic inspection, and annual inspection checklists are new features of the second edition. We hope these additions make your inspections easier and faster, and provide a higher degree of user safety.

IS THE CHECKLIST COMPREHENSIVE?

The *Safety First Checklist* is intended to provide a comprehensive list of potential safety hazards. However, due to limitations of current research information and the possibility of unique and unpredictable hazards, many critical inspection decisions should be made on-site by a trained playground inspector. In addition, some items were too variable to present in a brief checklist format. Some of these include harmful plants (e.g., local toxic species and trees that

INTRODUCTION *continued*

drop limbs) and former site uses (e.g., landfills or locations where hazardous chemicals have been stored or used).

Ongoing playground safety research will lead to improvements in the *Checklist* and ultimately in the design of play areas. For example, current areas of controversy, such as the use of signs, the use of chemical preservatives on play equipment, and site adjacencies to freeways and power lines will invariably lead to future revisions of the *Checklist*. As knowledge in the field of playground safety is always changing, the best protection against play area hazards continues to be a well-trained staff.

HOW OFTEN SHOULD PLAY AREAS BE INSPECTED?

Audit. An initial audit should be conducted to determine the current condition of the play area. Following this inspection, a list of hazards should be identified and corrective action should be taken. If it is necessary to remove, relocate, or replace play equipment, a design for play area renovation may be necessary.

Once the hazards identified in the audit are corrected, your play area will meet current safety guidelines. Your agency will then be ready to begin an ongoing risk management program of daily, periodic, and annual play area inspections.

Daily Inspection. A daily inspection should be conducted by maintenance or program staff to ensure that the play area is free of hazards, especially those due to vandalism.

Periodic Inspection. These inspections are conducted on a regular basis—monthly, bimonthly, or quarterly, depending on play area needs and conditions. This inspection assesses material wear and deterioration, such as wood rot, rust, worn hardware, or the need to replenish loose-fill safety surfacing. To determine how often to inspect a play area, the facility manager and risk manager should ask the following questions for each site (adapted from Kutska and Hoffman, 1992; Landscape Structures, 1989):

If the answer to any of the following questions is "Yes," a monthly inspection may be required:

1. Is loose-fill safety surfacing used in play equipment use zones?

2. Does play equipment have moving parts, such as chains, ring treks, track rides, etc.?

3. Does the site contain wooden play equipment or wooden site elements older than one year?

4. Does the site contain metal play equipment or metal site elements more than three years old?

5. Are severe climate conditions present, such as cold, hot, or wet weather, coastal climates, acidic soils, or poor drainage?

INTRODUCTION *continued*

6. Is park use high?

7. Is there a high rate of vandalism and documented repairs, or a greater than average number of reported accidents?

If the answer to the following questions is "Yes," a bimonthly or quarterly inspection may be required:

1. Is synthetic safety surfacing used in play equipment use zones?

2. Is play equipment free of moving parts, such as chains, ring treks, track rides, etc.?

3. Is wooden play equipment and/or site elements less than one year old?

4. Is metal play equipment and/or site elements less than three years old?

5. Is the climate mild and free of severe conditions, such as cold, hot, or wet weather, coastal climates, acidic soils, or poor drainage?

6. Is park use low to moderate?

7. Is there a low rate of vandalism and documented repairs, and few reported accidents?

Annual Inspection. The annual inspection is an in-depth evaluation of play area health and safety issues. In addition to completing the inspection, inspectors should review all hazards noted during the last year and corrective actions taken to ensure that all hazards have been promptly and correctly addressed.

HOW IS THE CHECKLIST SYSTEM ORGANIZED?

Audits, Annual Inspections, and Periodic Inspections. Separate checklist sections address the overall site, playground surfacing, and individual play equipment items (swings, climbers, etc.). The first page of each section includes an illustration or chart. Next, a series of "Yes" or "No" questions in checklist format allows the inspector to assess safety factors. Each checklist section includes audit questions, annual inspection questions, and periodic inspection questions.

For audits: the audit, annual inspection, and periodic inspection questions should be completed.

For annual inspections: the annual and periodic inspection questions should be completed.

For periodic inspections: the periodic inspection questions should be completed.

For all inspection questions: "No" answers indicate a potential safety hazard. If the answer to any item on the checklist is "No," removal or repair of equipment and/or a site element may be necessary. Play area redesign may also be needed.

INTRODUCTION *continued*

Daily Inspections. All inspections questions needed for the entire play area are included in the Daily Inspection Checklist. Again, "No" answers to checklist questions indicate a potential safety hazard.

HOW SHOULD THE CHECKLIST BE USED?

Before you begin your inspection, follow the steps described below:

1. Review the Definitions section beginning on page 7. This section defines important terms used throughout the *Checklist*. An understanding of these terms is necessary to perform a play area inspection.

2. Assemble the needed inspection tools, described on pages 14 and 15. These tools will be needed when you inspect your play areas.

3. Review and practice the inspection procedures described beginning on page 16. These procedures will enable you to assess playground equipment for potential head and neck entrapments, protrusions, and accessible pinch, crush, or shear points.

Audits, Annual Inspections, and Periodic Inspections. To complete an audit, annual inspection, or periodic inspection of your play area, follow these steps:

1. Each audit, annual inspection, or periodic inspection begins with an overall site assessment using the Site Survey checklist (#1). For audits or when changes have occurred to the play area layout, draw a simple sketch of the site layout in the space provided on the front of this section. If play area use zones are inadequate, indicate the distance between play equipment on your sketch. During the audit, the dimensions of the play area should be noted in the space provided. These dimensions will be helpful in calculating costs if redesign becomes necessary. Next, complete the checklist questions indicated for the type of inspection you are performing (audit, annual, or periodic).

2. Complete the Safety Surfacing checklist (#2). This section includes inspection questions for synthetic surfacing, wood-product surfacing, sand surfacing, gravel surfacing, and chopped-tire surfacing. For each type of surfacing used in your play area, complete the checklist questions indicated for audits, annual inspections, or periodic inspections. If the play area has no impact-attenuating surfacing in play equipment use zones, it should be closed immediately.

3. After assessing the overall site and surfacing conditions, evaluate each piece of equipment in the play area. Select the appropriate checklist for each item (e.g., swings, climbers, etc.). Next, complete the checklist questions indicated for audits, annual inspections, or periodic inspections. If during the course of the inspection it becomes evident that a particular

INTRODUCTION *continued*

piece of equipment must be removed, you may wish to stop the inspection of that equipment.

Example 1. A horizontal ladder does not have the appropriate unobstructed use zone. Field inspections showed that it was free of severe structural deterioration and did not exceed the recommended equipment height. The inspector's conclusion was that this structure needed relocation. In this case, the play area inspector continued the survey to assess whether the horizontal ladder would be safe if relocated to a 72-inch use zone.

Example 2. During the survey, a composite structure was found to have severe wood rot in all vertical supports. The structure was unstable. It was evident that the structure could not be repaired and must be immediately removed. After discovering its poor condition, the inspector discontinued the evaluation of this equipment.

4. If the play equipment includes ladders, stairways, ramps, stepped platforms, or transfer points, the Equipment Access & Egress checklist (#3) should be attached to the equipment checklist and used during the audit. Both the general considerations and the specific questions for each type of access provided should be completed. During annual or periodic inspections, the Equipment Access & Egress checklist will not be needed. The condition of access components can be determined using the checklist for the equipment type provided.

5. If you are evaluating equipment that includes platforms requiring protective barriers, attach the Guardrails & Protective Barriers checklist (#4) to the equipment checklist. This checklist can also be used to inspect freestanding game panels. Complete the checklist questions indicated for the type of inspection you are performing (audit, annual, or periodic).

6. Evaluate composite structures by completing the Composite Structures checklist (#22) and a checklist for each type of play equipment attached to the structure. For example, to evaluate a composite structure that includes a horizontal ladder and a rotating swing, select the following checklists: Composite Structures, Horizontal Ladders & Ring Treks, Rotating Swings, Guardrails & Protective Barriers, and Equipment Access & Egress. Next, complete the checklist questions indicated for the type of inspection you are performing (audit, annual, or periodic).

7. Site-built and community-built structures are common in every jurisdiction. To evaluate these structures, select the checklist section or the combination of sections that best describes the structure. Next, complete the checklist questions indicated for the type of inspection you are performing (audit, annual, or periodic).

Daily Inspections. For daily inspections, complete the Daily Inspection Checklist on page 22.

INTRODUCTION *continued*

HOW SHOULD HAZARDS BE REPORTED?

As you perform the inspection, all checklist questions answered "No" indicate a hazard. As problems are noted, one or more of the following actions should be taken:

1. Perform a maintenance task to correct the hazard; or

2. Close the hazardous play equipment (or the entire play area) and submit a work order.

It is important to document all hazards found during an inspection and all corrective actions taken. A sample form that can be used to report inspection results has been included at the back of this book. Completed inspection checklists and inspection summary reports should be kept in a file to document your inspection program.

HOW SHOULD PLAY AREA ACCESSIBILITY BE EVALUATED?

The first edition of the *Safety First Checklist,* published in 1989, included a pioneering effort to address access for people with disabilities in public parks and play areas. Since then, the *Americans with Disabilities Act Accessibility Guidelines* (ADAAG) have been adopted. The American Society for Testing and Materials (ASTM) has also addressed accessibility issues in play areas in versions of ASTM F 1487 published in 1993 and 1995. In addition, draft guidelines for both outdoor recreational environments and children's facilities are now being considered for inclusion in ADAAG by the U.S. Architectural and Transportation Barriers Compliance Board. When adopted, these guidelines will provide further clarification for public recreation agencies.

These efforts are the result of years of work by people with disabilities and advocates to encourage the development of play areas that provide a range of play opportunities and experiences for people with varying disabilities in an integrated social setting. While the focus of this manual is on safety rather than access, we fully support this concept.

This edition of *Safety First Checklist* addresses access only as it might be considered during a safety inspection, providing inspection checklists for wheelchair ramps and transfer points. To assist public recreation agencies in fully evaluating park and recreation facility access, MIG Communications has published a separate document to fully address accessibility issues. *The Accessibility Checklist* (Goltsman, Gilbert, and Wohlford, 1993) can be used to evaluate your agency's compliance with the Americans with Disabilities Act (ADA), and will be updated to reflect new guidelines as they become available. For more information, call MIG Communications at (510) 845-0953.

DEFINITIONS

clearance zone. A clearance zone is a required unobstructed zone surrounding play equipment.

slide clearance zone. This zone (Figure 1) should extend at least 60 inches (1500 mm) above the chute surface and 21 inches (530 mm) from each side (measured from the inside face of the walls). The zone should extend through the slide exit region. Slide hoods, guardrails, or other devices intended to channel users into a sitting position are excluded from this guideline. For spiral slides, the zone should extend 21 inches (530 mm) from the inside face of the outer edge of the slide along the entire length of the slide.

rotating swing clearance zone. The minimum clearance zone (Figure 2) should be a cylindrical area with a radius equal to the distance from the pivot point to the sitting surface of the seat plus 30 inches (760 mm). This distance should extend both sides of the pivot point. The vertical height of the clearance zone should extend from the top of the safety surface to the height of the pivot point throughout the horizontal length of the clearance zone.

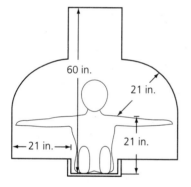

Figure 1. Slide clearance zone.

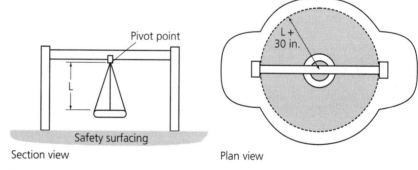

Figure 2. Rotating swing clearance zone.

composite structure. A composite structure consists of two or more play events that are attached or functionally linked to create one integral unit.

crush points. See *pinch, crush, and shear points*.

entanglement. Entanglement occurs when a person's clothing or items worn around the person's neck become caught or entwined on play equipment. Entanglement can result in strangulation, loss of a body part, or emotional injury. To prevent entanglement, fasteners should be closed (see *fasteners and connecting devices*), and protrusions should meet ASTM F 1487 requirements (see Inspection Procedures).

DEFINITIONS *continued*

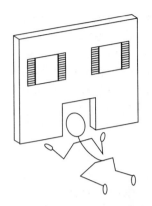

Figure 3. Openings with surfacing serving as the bottom edge are exempt from entrapment criteria.

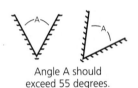

Angle A should exceed 55 degrees.

Angle A is exempt if one of its legs is horizontal or slopes downward from the apex.

Figure 4. Recommendations for angular openings.

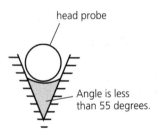

Figure 5. Angle with a filled apex that does not contact both sides of the probe.

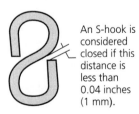

An S-hook is considered closed if this distance is less than 0.04 inches (1 mm).

Figure 6. Completely closed fastening device.

entrapment. Entrapment occurs when a child's head or limb becomes lodged within a space and cannot be withdrawn. Entrapment can result in strangulation, loss of fingers or limbs, or emotional injury. Openings that are accessible to children should meet all entrapment criteria specified in ASTM F 1487. Both the size and shape of the opening are considered when determining entrapment potential. See Inspection Procedures for a description of tests used to identify openings that may be an entrapment.

accessible completely bounded openings. An opening is considered accessible when a torso probe (see Inspection Tools) can be inserted into the opening to a depth of 4 inches (100 mm) or more. Openings with the ground surface serving as the lower edge of the opening are exempt (Figure 3).

rigid completely bounded openings. Rigid openings have sides that are fixed and do not change form, such as the space between platforms, ladder rungs, and steps.

nonrigid completely bounded openings. Nonrigid openings change shape when pressure is applied, such as flexible nets, tarps, and plastic enclosures.

completely bounded openings with limited depth. Openings with depth limited by barriers contain two potential entrapment areas: a vertical opening and a horizontal opening (see Inspection Procedures).

partially bounded openings. Any opening in a piece of play equipment that is not totally enclosed by boundaries on all sides. Openings that are inverted (i.e., the lowest interior boundary next to the opening is horizontal or slopes downward) are exempt from entrapment inspection requirements. In addition, openings that allow full passage of the head probe are not considered an entrapment (see Inspection Procedures).

accessible partially bounded opening. A partially bounded opening is considered accessible if the unbounded portion is between 1.875 (48 mm) and 9 inches (228 mm).

angular openings. Angular openings are formed by intersecting surfaces or surfaces that would intersect if projected. All angular openings should measure at least 55 degrees (Figure 4). The distance between angle surfaces should be greater than 9 inches (228 mm) to prevent head entrapment. An angle is exempt from these head entrapment requirements if the lower edge of the angle is horizontal to the ground or slopes downward. An angle is also exempt if its apex is filled so that the 9-inch (228 mm) head probe cannot touch both sides of the angle simultaneously when the probe is rotated in any direction (Figure 5).

fasteners and connecting devices. Fasteners and connecting devices discourage unintentional loosening and reduce friction between moving elements and fixed supports. Fastening devices such as S-hooks, pelican

DEFINITIONS *continued*

This S-hook fails because the bottom leg extends past the boundary lines established by the top of the exposed loop of the S-hook.

Figure 7. End wire of S-hook extends beyond upper loop.

hooks, C-hooks, and clevis devices should be closed to prevent entanglement. Fastening devices are considered closed when the space between points measures less than 0.04 inches (1 mm) (Figure 6). When an S-hook is closed, the lower loop should not extend beyond the upper loop more than 1/8 inch (3 mm) (Figure 7). Connecting devices should not spin and create an entanglement. Fasteners and connecting devices should consist of corrosion-resistant material, such as stainless steel, brass, zinc-plated metal, zinc-chromate-plated metal, or galvanized steel.

guardrails. A guardrail is a partially enclosed barrier used to help prevent children from falling off elevated platforms. CPSC and ASTM allow the use of guardrails for some platform heights and age groups. Guardrails, however, provide less protection. ASTM F 1487 requires guardrails for elevated surfaces that are greater than 20 inches (510 mm) when used for 2- to 5-year-olds, and on elevated surfaces greater than 30 inches (760 mm) when intended for use by 5- to 12-year-olds. ASTM requires protective barriers for elevated surfaces greater than 30 inches (760 mm) for 2- to 5-year-olds and 48 inches (1200 mm) when intended for use by 5- to 12-year-olds. Figure 8 illustrates ASTM requirements for guardrails.

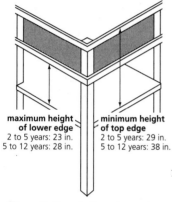

maximum height of lower edge
2 to 5 years: 23 in.
5 to 12 years: 28 in.

minimum height of top edge
2 to 5 years: 29 in.
5 to 12 years: 38 in.

Figure 8. Guardrail requirements. Height represents the distance above the platform.

maximum equipment heights. A common hazard in children's play areas is play equipment that is inappropriate for the users' age group. Equipment should allow safe and successful use by children of a specific chronological age, mental age, and physical ability. Selected play equipment should be of an appropriate height and complexity for the age group of the intended users. ASTM F 1487 provides guidance for maximum equipment heights for several types of equipment. These recommended heights are included in the appropriate checklist. An individual child's skills, however, may vary from these averages. In such cases, play ability should be assessed by parents, guardians, and staff.

maximum user. The maximum user of play equipment is a 12-year-old in the 95th percentile, approximately 62 inches (1600 mm) tall and 120 pounds (55 kg).

paint. All paints and similar finishes must comply with ASTM F 1487 requirements to minimize lead exposure (0.06% maximum lead by dry weight). Manufacturers should verify that paints meet this specification.

pinch, crush, and shear points. These junctures can cause contusion, laceration, abrasion, amputation, or fracture during use. Defined as any point that entraps a 5/8-inch (16 mm) diameter rod at one or more positions, these points are created when components move in relationship to each other or to a fixed component (see Inspection Procedures). Chains and the attachment area of heavy-duty coil springs to the base of rocking equipment are exempt.

©1997 MIG Communications

DEFINITIONS *continued*

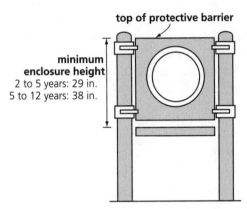

Figure 9. Protective barrier requirements.

protective barriers. Protective barriers are enclosures that help prevent children from falling off elevated platforms. These barriers can consist of a vertical surface, play event, series of vertical or diagonal bars spaced less than 3½ inches (89 mm) apart, or other barriers free of footholds or handholds that would encourage climbing (Figure 9).

protective barriers for 2- to 5-year-olds. Any play equipment platform more than 30 inches (760 mm) above the ground must be enclosed by a protective barrier at least 29 inches (740 mm) high.

protective barriers for 5- to 12-year-olds. Any play equipment platform more than 48 inches (1200 mm) above the ground must be enclosed by a protective barrier at least 38 inches (970 mm) high.

protrusions. Protrusions are components that extend in any direction from play equipment, site elements, or site furnishings. Protrusions can catch a child's clothing, causing strangulation or loss of balance. Protrusions also pose a potential impact hazard. Protrusions must meet ASTM F 1487 requirements (see Inspection Procedures).

safety surfacing. Safety surfacing consists of shock-absorbent surfaces that must be used throughout play equipment use zones. Safety surfaces include synthetic materials (e.g., poured-in-place surfacing, tiles) and loose-fill materials (e.g., wood products, sand, gravel, chopped tire).

Safety surfaces should meet the impact-attenuating requirements of CPSC when tested in accordance with ASTM F 1292. These requirements were developed to reduce the likelihood of life-threatening head injuries caused by falls from play equipment. The surfacing must ensure that a head-first fall from the highest accessible height of the equipment onto the play area surface will result in an impact that does not exceed 200 g's and a Head Injury Criteria (HIC) value of no more than 1,000, as specified by ASTM F 1292. (The 200-g's requirement refers to the *g-force*, the force of gravity on an object. G-force measures the peak deceleration of the head during impact. It takes the weight of the object and the distance of fall into consideration. HIC measures the duration of the impact during its most severe phase.) The highest accessible height of various types of play equipment is measured from different points:

Play equipment	Highest accessible height
composite equipment	measure from the height of the highest platform when enclosed by protective barrier; measure from top of guardrail when enclosed by guardrail
spring rockers	measure from the seat
stationary, climbable equipment	measure from the maximum height of the structure
swings	measure from the height of the swing pivot point

DEFINITIONS *continued*

sharp points, corners, and edges. Sharp edges may cut or puncture a person's skin and should be avoided in play areas, as prescribed by ASTM F 1487. Exposed open ends of tubing should be covered with caps or plugs that cannot be removed without tools. Bolt ends should not extend more than two threads beyond the face of the nut, and should be free of burrs and sharp edges. The corners and edges of suspended parts should have a minimum curvature radius of 1/4 inch (6 mm). Flexible components such as belts, straps, and ropes are exempt.

shear points. See *pinch, crush, and shear points*.

slide height/length ratio. The height of the slide bed divided by its length should not exceed 0.577 (Figure 10).

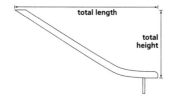

Figure 10. Height of slide divided by length should not exceed 0.577.

structural integrity. ASTM has developed test procedures that measure the load-bearing capacity of manufactured play equipment. The manufacturer should verify that the play equipment has been tested and meets all standards for structural integrity, as specified by ASTM F 1487.

suspended hazards. Hazards can be created by chains, cables, or ropes suspended between play equipment or from play equipment to the ground. Flexible elements should not be suspended within 45 degrees of horizontal at a height less than 84 inches (2100 mm) above the playground surface. Suspended elements must measure at least 1 inch (25 mm) in diameter at the narrowest cross section. They should be fixed at both ends, and should not be capable of being looped back in itself. Two or more suspended cables or wires that are located at two or more heights may be suspended below 84 inches (2100 mm) if they meet all of the above requirements and cannot be looped or stretched to contact another suspended component. Suspended elements should be brightly colored or contrast with surrounding equipment to increase visibility.

transfer point. A platform along an accessible route of travel that allows a wheelchair user to transfer from the wheelchair onto play equipment. A transfer point and adjacent platforms or steps that allow a child to move through the equipment is called the transfer system.

use zone. A use zone is an obstacle-free area under and around play equipment where a child could land when falling from, jumping from, or exiting the equipment. The entire use zone must be covered with safety surfacing.

use zone dimensions. Use zone dimensions depend on the equipment type. ASTM F 1487 recommends a minimum use zone for typical stationary equipment of 72 inches (1800 mm) extending from all sides of the equipment (Figure 11, page 12). Slide and swing use zones have other requirements, which are described on pages 12 and 13.

overlapping use zone. Use zones generally may not overlap. Two pieces of stationary play equipment that are functionally linked, such as two balance beams or two nonclimbable playhouses, and that have no designated play surfaces higher than 30 inches (760 mm) may have overlapping use zones (Figure 12, page 12). These exceptions are noted on individual checklists.

DEFINITIONS *continued*

Two play elements that must have nonoverlapping use zones cannot be closer than the distance obtained by adding together both required use zones.

supplemental circulation area. Sufficient space must be provided between all adjacent structures and play equipment for circulation and for play. In settings where periodic overcrowding is likely, a supplemental circulation area beyond the use zone is recommended.

swing use zone. This zone (Figure 13) should extend in front and back of the swing crossbeam a distance measuring at least twice the height of the pivot point. The zone should also extend at least 72 inches (1800 mm) from both sides of the support structure. Adjacent swing structures can share the 72-inch (1800 mm) use zone at the side.

swing use zone (enclosed seats). Because users of swings with enclosed seats, such as tot seats and bucket seats, cannot intentionally exit from the seats, a somewhat smaller use zone may be provided. This zone should extend both in front and back of the swing crossbeam a distance equal to at least

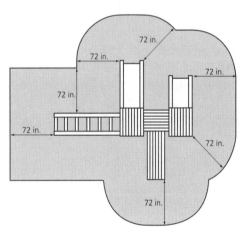

Figure 11. Typical stationary equipment use zone.

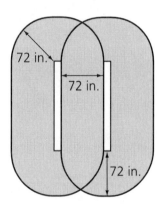

Figure 12. Overlapping use zones.

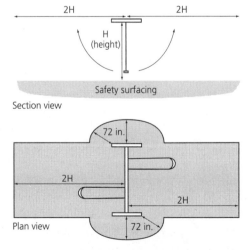

Figure 13. Swing use zone.

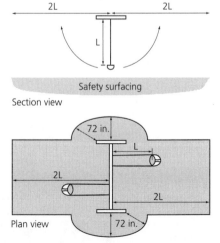

Figure 14. Enclosed-seat swing use zone.

DEFINITIONS *continued*

twice the distance measured from the top of the occupant's sitting surface to the swing pivot point (Figure 14). The zone should also extend at least 72 inches (1800 mm) from both sides of the support structure. Adjacent swing structures can share the 72-inch (1800 mm) use zone at the side.

rotating swing use zone. This zone (Figure 15) should extend in all directions from the pivot point of the swing. The distance should measure at least 72 inches (1800 mm) plus the vertical distance between the pivot point and the top of the swing seat. The zone should also extend at least 72 inches (1800 mm) from both sides of the support structure. Adjacent swing structures can share the 72-inch (1800 mm) use zone on the side.

slide use zone. The slide use zone (Figure 16) should extend at least:

- 72 inches (1800 mm) from all sides of the entry steps or platform;
- 72 inches (1800 mm) from both sides of the slide chute; and
- at the slide exit zone, a distance equal to the height of the slide entrance zone plus 48 inches (1200 mm), measured from the point where the slide's slope decreases to less than 5 degrees from the horizontal. A minimum use zone of 72 inches (1800 mm) from the slide exit zone must be provided.

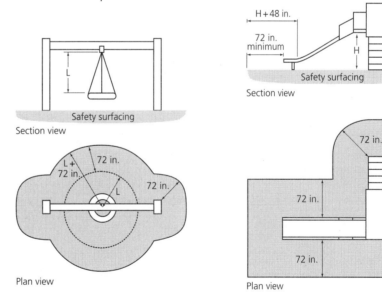

Figure 15. Rotating swing use zone.

Figure 16. Slide use zone.

wood preservatives. Wood used in playground equipment should be naturally rot- and insect-resistant or treated with a wood preservative that meets ASTM F 1487 requirements. Chromated copper arsenate (CCA) is acceptable for use if the dislodgeable arsenic on the wood surface is minimized. However, arsenic-treated wood should not be used to construct drinking fountains or other water sources. Copper or zinc naphthenates and borates have low toxicity and are suitable for use in play areas. Creosote, pentachlorophenol, and tributyl tin oxide should not be used. Pesticide-containing finishes should also not be used. Manufacturers should provide verification that wood preservatives used do not pose a hazard to play area users.

INSPECTION TOOLS

The tools needed for play area inspections are available commercially (except for the test template for partially bounded openings, which you will need to construct yourself). Inspection tools should be kept in a tote bag for easy access and portability during inspections.

Play area inspection tools include:

- 25-foot (8 m) tape measure
- electronic level for measuring slopes
- 24-inch (600 mm) ruler
- torso probe* (Figure 17)
- head probe* (Figure 18)
- articulated web stop probe† (Figure 19)
- set of three protrusion inspection gauges* (Figure 20)
- protrusion gauge for swing seats and hardware* (Figure 21)
- test template for partially bounded openings (Figure 22) (construct from any rigid material 0.75 in (19 mm) thick)

* CPSC- and ASTM-recommended tools available from:
National Recreation and Park Association (NRPA)
2775 South Quincy St.
Suite 300
Alexandria, VA 22206
(703) 820-4940

† ASTM-recommended tool available from:
Underwriters Laboratories
333 Pfinsten Rd.
Northbrook, IL 60062
(847) 272-8800 ext. 42612

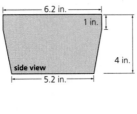
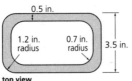

Figure 17. Torso probe.

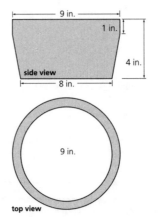

Figure 18. Head probe.

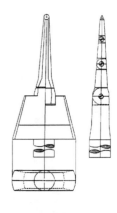

Figure 19. Articulated web stop probe.

INSPECTION TOOLS *continued*

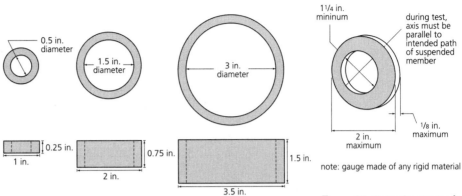

Figure 20. Set of three protrusion inspection gauges.

Figure 21. Protrusion gauge for swing seats and swing hardware.

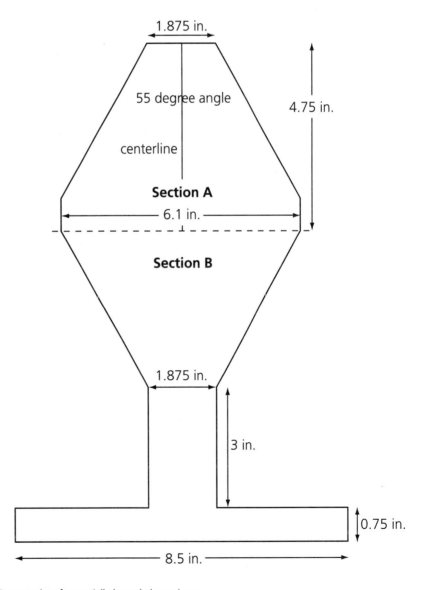

Figure 22. Test template for partially bounded openings.

INSPECTION PROCEDURES

HEAD AND NECK ENTRAPMENT INSPECTION: COMPLETELY BOUNDED OPENINGS

Any completely bounded opening that is accessible to children should meet all entrapment inspection requirements. The torso probe and head probe (Figures 17 and 18, page 14) should be used to inspect these openings for possible head and neck entrapments. Openings between the safety surfacing and the bottom edge of the play area element (Figure 3, page 8) are exempt from entrapment requirements.

completely bounded openings. Any opening enclosed on all sides with a continuous perimeter is considered a completely bounded opening.

accessible openings. A completely bounded opening is considered accessible when a torso probe can be inserted into the opening to a depth of 4 inches (100 mm) or more.

Inspecting rigid openings

Rigid openings, such as openings between platforms, rung ladders, and steps, are fixed and do not change form. To pass the inspection, the opening must not permit the torso probe to pass through it, *or* must permit *both* the torso probe and head probe to pass through it.

1. *Using the torso probe.* Hold the torso probe parallel to the opening and attempt to insert the probe. If the probe does not fit through the opening to a depth of 4 inches (100 mm) or more when rotated in any direction, the opening is not a potential entrapment and does not require further inspection. If the torso probe does pass through the opening, the opening must be inspected with the head probe. For protective barriers and game panels functioning as barriers, openings within the barrier and between the barrier's lower edge and platform surface should preclude the passage of the torso probe.

2. *Using the head probe.* Hold the head probe parallel to the opening and attempt to insert it through the opening. If the probe does not pass freely through the opening, the space is a potential entrapment.

Inspecting nonrigid openings

Nonrigid openings, such as openings in flexible nets, tarps, and plastic enclosures, change shape when pressure is applied. To pass the inspection, the opening must not permit the base of the torso probe to pass through it, *or* must permit *both* the torso probe and head probe to pass through it.

INSPECTION PROCEDURES continued

1. *Using the torso probe.* Hold the torso probe parallel to the opening. Attempt to push or pull the probe through the opening using no more than 50 lbf (222 N) of pressure. If the base of the probe does not fit through the opening when rotated in any direction, the opening is not a potential entrapment and does not require further inspection. If the torso probe does pass through the opening, the opening must be inspected with the head probe.

2. *Using the head probe.* Hold the head probe parallel to the opening and attempt to insert it through the opening. If the probe does not pass freely through the opening, the space is a potential entrapment.

Inspecting openings with limited depth

An example of an opening with limited depth is a ladder with a barrier behind it. In openings with limited depth (Figure 23), there are two potential entrapment areas: a vertical opening (A) and a horizontal opening (B). The inspection procedure emulates a child crawling into the vertical opening feet first and passing downward through the horizontal space. To pass the inspection, the vertical opening (A) must not permit the torso probe to pass through it, *or* the torso probe may pass through the vertical opening (A) but not the horizontal opening (B). If the torso probe passes through both openings, the head probe must also pass through both openings.

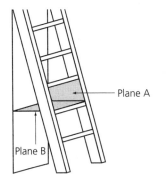
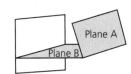

Figure 23. Opening with limited depth.

1. *Inspecting the vertical opening (A) with the torso probe.* Hold the torso probe parallel to the opening and attempt to insert it through the opening. If the probe does not fit into the opening when rotated in any direction, the opening is not a potential entrapment and does not require further inspection. If the torso probe does pass through the opening, the horizontal opening (B) must be inspected with the torso probe.

2. *Inspecting the horizontal opening (B) with the torso probe.* Hold the torso probe horizontally with the longest end of the tool against the edge of the vertical opening (A). Attempt to insert the probe through the opening. If the probe does not fit through the opening, the space is not large enough for a child to completely enter it and is not a potential entrapment. If the torso probe does pass through the horizontal opening (B), both openings (A and B) must be inspected with the head probe.

3. *Inspecting the vertical and horizontal openings (A and B) with the head probe.* Hold the head probe parallel to both openings. Attempt to insert it through the openings. If the head probe passes through both spaces, there is no potential entrapment. If the probe does not pass freely through both openings, the space is a potential entrapment.

INSPECTION PROCEDURES *continued*

HEAD AND NECK ENTRAPMENTS: PARTIALLY BOUNDED OPENINGS

Any partially bounded opening that is accessible to children should meet entrapment inspection requirements. The test template for partially bounded openings (Figure 22, page 15), the head probe (Figure 18, page 14), and a tape measure should be used to inspect openings for possible head and neck entrapment. Openings that are inverted are exempt from these requirements. A partially bounded opening is considered inverted if the lowest interior edge of the opening is horizontal or slopes downward (Figure 24). In addition, openings that allow full passage of the head probe are not considered an entrapment.

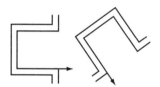

Figure 24. Inverted openings, in which the lowest edge of the opening is either horizontal or sloping downward, are exempt from entrapment requirements.

partially bounded opening. Any opening in a piece of play equipment that is not totally enclosed by boundaries on all sides.

accessible partially bounded opening. The opening is considered accessible if the unbounded portion is between 1.875 (48 mm) and 9 inches (228 mm).

1. *Inspecting the opening with Section A of the test template.* To inspect a partially bounded opening, you must first test it with Section A of the test template. This test determines whether the opening is large enough to potentially entrap the head of a 5-year-old or the neck of a 2-year-old. Section A of the test template should be inserted into the opening until the template makes contact with the opening's boundaries (Figure 25). The centerline of the template should be aligned with the centerline of the opening. If the sides of the template make simultaneous contact with the sides of the opening, the opening fails and must be tested using Section B of the test template. If there is not simultaneous contact, however, the opening passes and the inspection is complete.

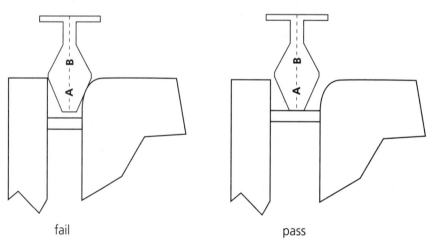

Figure 25. Partially bounded openings (Section A). If the sides of the template make simultaneous contact with the sides of the opening, the opening fails and must be tested using Section B.

2. *Inspecting the opening with Section B of the test template.* This test determines whether the opening shape and depth could entrap a child's neck. Hold Section B of the test template perpendicular to the opening (Figure 26). Insert the template into the opening. If the test template fits com-

INSPECTION PROCEDURES *continued*

pletely within the opening, the opening is deep enough to be hazardous and fails the test unless the opening fully allows the head probe to pass through it. If the test template does not fit completely inside the opening, the opening passes the test and is not a hazard.

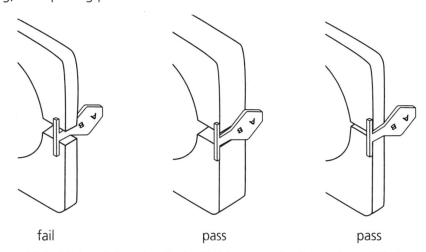

fail   pass   pass

Figure 26. Partially bounded openings (Section B). If the test template fits completely within the opening, and the opening does not allow the head probe to pass through it completely, the opening is a hazard.

HEAD AND NECK ENTRAPMENT INSPECTION: ANGULAR OPENINGS

Angles may be formed by adjacent intersecting surfaces or by surfaces that would intersect if projected. All angles formed by the surfaces of an opening should measure at least 55 degrees, unless it meets exemption requirements (see Definitions under *entrapment*) (Figure 4, page 8). Use the head probe to inspect angular openings for potential entrapment. The angle passes the inspection if the head probe cannot simultaneously contact both sides of the angle as follows:

1. *Inspecting angular openings with the head probe.* The distance between angle surfaces should be greater than 9 inches (228 mm) to prevent head entrapment. To measure compliance, insert the head probe between the angle surfaces. If the head probe cannot contact both surfaces of the angle simultaneously when the probe is rotated to any orientation, the angle is not a potential entrapment.

2. *Inspecting inverted angles.* An angle is inverted if the lower edge of the angle is horizontal or slopes downward (Figure 4, page 8). An inverted angle cannot entrap the head or neck, and is exempt from requirements for angular openings.

3. *Inspecting angles with a filled apex.* To measure compliance, insert the head probe between the angle surfaces. If an angle less than 55 degrees is infilled so that the head probe cannot contact both surfaces of the angle simultaneously when the probe is rotated to any orientation, the angle is not a potential entrapment (Figure 5, page 8).

INSPECTION PROCEDURES *continued*

PROTRUSION INSPECTION

Hardware, pipes, posts, or other structural elements that extend in any direction from play equipment should be inspected with the set of three protrusion gauges (Figure 20, page 15). A separate protrusion gauge is used to inspect swing seat and swing hardware protrusions (Figure 21, page 15). A measuring tape or ruler is also required. A protrusion is considered inaccessible and exempt from protrusion requirements when it is recessed or located so that a protrusion gauge cannot be placed over it. Requirements for passing the inspection vary with the type of protrusion, as described below:

1. *Inspecting protrusions.* Protrusions should be tested by successively placing each of the three test gauges over the protrusion. If the protrusion fits inside any gauge, the protrusion cannot extend beyond the face of the gauge to pass the inspection. Each protrusion should be visually inspected to ensure that no protrusion increases in diameter from the surface to the exposed end. Any caps or coverings should also be visually inspected to ensure that they fit flush against the nut or surrounding surface.

2. *Inspecting compound protrusions.* For compound protrusions, successively place each of the three test gauges over increasing diameters to determine compliance (Figure 27). To pass the inspection, protrusions must not extend beyond the face of any of the three gauges.

3. *Inspecting vertical protrusions.* If a vertical protrusion fits within any of the three test gauges, the length of the protrusion should be measured. To pass the inspection, the protrusion must not project more than 1/8 inch (3 mm) above the adjacent horizontal surface.

4. *Inspecting compound vertical protrusions.* For compound vertical protrusions, the length of each individual protrusion surface should be measured (Figure 28). To pass the inspection, the length of each individual protrusion surface cannot project more than 1/8 inch (3 mm) above the adjacent horizontal surface. Next, the compound vertical protrusions should be tested by successively placing each of the three test gauges over the pro-

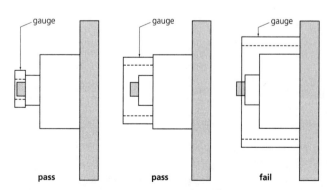

Figure 27. Compound protrusions. Successively place gauges over increasing diameters.

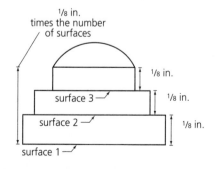

Figure 28. Compound vertical protrusions. After passing the regular protrusion test, no vertical protrusion should project more than 1/8 inches above the previous horizontal surface.

INSPECTION PROCEDURES *continued*

trusion. To pass the inspection, the total length of the protrusion must not extend beyond the face of any of the three test gauges.

5. *Inspecting swing seat and swing hardware protrusions.* The swing protrusion test gauge (Figure 21, page 15) is used to test protrusions on the front or rear surfaces of suspended swings. The gauge should be placed over the protrusion. To pass the inspection, the protrusion must not extend beyond the face of the gauge.

PINCH, CRUSH, AND SHEAR POINT INSPECTION

Openings that may provide access to potential pinch, crush, and shear points should be inspected with the articulated web stop probe (Figure 19, page 14). A tape measure will also be needed. Requirements for passing the inspection are described below:

1. *Inspecting openings with a minimum width of less than 1 inch (25 mm).* Insert the articulated web stop probe point first into the opening in all possible positions with a force that does not exceed 1 pound (4 N). To pass the inspection, the probe must not touch any pinch, crush, or shear point.

2. *Inspecting openings with a minimum width of 1 inch (25 mm) or more.* When potential pinch, crush, or shear points are covered with material that contains openings of 1 inch (25 mm) or more, measure the distance from the opening to the potential pinch, crush, or shear point. To pass the inspection, the opening must meet the following requirements:

Minimum width of opening		Minimum distance from opening to part	
inches	mm	inches	mm
1	25	6 1/2	165
1 1/4	32	7 1/2	190
1 1/2	38	12 1/2	320
1 7/8	48	15 1/2	395
2 1/8	54	17 1/2	445

DAILY INSPECTION CHECKLIST

PARK NAME DATE OF INSPECTION INSPECTOR

Note: CPSC and ASTM F 1487 do not address general site safety. The questions below reflect the opinion of the authors.

Yes	No	N/A		
☐	☐	☐	1.	Is the play area free of trash?
☐	☐	☐	2.	Are the trash receptacles and surrounding area free of spills?
☐	☐	☐	3.	Is the play area free of mushrooms and other fungi?
☐	☐	☐	4.	Is the play area free of known toxic plants and plants with thorns?
☐	☐	☐	5.	Have all broken or sharp branches been removed from trees and shrubs?
☐	☐	☐	6.	Are all irrigation heads retracted?
☐	☐	☐	7.	Are all sinks and drinking fountains clean, unclogged, and functioning?
☐	☐	☐	8.	Are all storm drains unclogged?
☐	☐	☐	9.	Is the play area free of insect infestation?
☐	☐	☐	10.	Is the play area free of rodents and other vertebrate pests?
			11.	When loose-fill safety surfacing is used:
☐	☐	☐		a. Is the surfacing free of low areas and holes?
☐	☐	☐		b. Is the surfacing free of animal feces and hidden debris?
☐	☐	☐		c. Is the surfacing completely unfrozen when the play area is in use?
			12.	When synthetic safety surfacing is used:
☐	☐	☐		a. Is the outdoor temperature more than 25°F (12°C) when the play area is in use?
☐	☐	☐		b. Is the surfacing free of slippery materials, such as sand, gravel, leaves, soil, puddles, ice, or snow?
☐	☐	☐	13.	Are hard surfaced pathways free of slippery materials, such as sand, ice, snow, algae, or moss?
☐	☐	☐	14.	Are play equipment platforms, stairs, or rungs free of slippery materials, such as water, ice, sand, or leaves?
☐	☐	☐	15.	Is the play area free of loose, missing, or broken parts?
☐	☐	☐	16.	Is the play area free of sharp edges?
☐	☐	☐	17.	Are there no metal equipment surfaces that are excessively hot or cold to the touch when the play area is in use?
☐	☐	☐	18.	Has all scrap material added to the play equipment been removed, such as ropes, sheets, or cardboard?

©1997 MIG Communications

1 SITE SURVEY

Note: CPSC and ASTM F 1487 do not address general site safety. The questions in this checklist reflect the opinion of the authors.

Park name _____

Owner _____

Address _____

Date of inspection _____

Inspector _____

Supervised hours _____ **Unsupervised hours** _____

Number of hours of maintenance per week _____

Dimensions of play area _____

Sketch the layout of the play area below, indicating the location of play equipment.
Scale: 1 inch = _____ feet (4 squares per inch)

SITE SURVEY

| PARK NAME | DATE OF INSPECTION | INSPECTOR |

Yes	No	N/A		**AUDIT**
☐	☐	☐	1.	Is the play area located away from land uses that are incompatible with children's play? ▷ *Play areas should be located away from sources of loud noise, air pollution, or high traffic volume; air fields; railroads; former landfills; industrial areas; and other areas where hazardous materials are used.*
☐	☐	☐	2.	Is the play area located away from electrical lines and transformers? ▷ *An electromagnetic field specialist should be consulted to determine requirements.*
☐	☐	☐	3.	Is safe access to the play area provided for pedestrians and bicyclists?
☐	☐	☐	4.	Is access provided for maintenance, emergency, and service vehicles?
☐	☐	☐	5.	Are automobile parking and driving areas separated from pedestrian and waiting zones?
☐	☐	☐	6.	Is play equipment designed for children under 5 years separated from equipment designed for 5- to 12-year-olds?
☐	☐	☐	7.	Are storm drains located outside the play area? ▷ *Drainage grates should never be located in play equipment use zones unless the grates are covered with synthetic impact-attenuating surfacing.*
☐	☐	☐	8.	Are all utility boxes located outside of the play area? ▷ *Utility boxes should never be located in play equipment use zones.*
☐	☐	☐	9.	Are provisions for utility metering, transformers, and other electrical equipment located in locked vaults or utility rooms away from children's play areas?
☐	☐	☐	10.	Is a telephone provided for emergency communication?
☐	☐	☐	11.	Is perimeter fencing provided for play areas serving children under 5 years and for all play areas when required by traffic volume?
☐	☐	☐	12.	Are informational signs provided about the use of the park and equipment, including the intended age group of users?
☐	☐	☐	13.	Is an appropriate use zone provided for all play equipment? ▷ *See individual checklists for each type of equipment.*
☐	☐	☐	14.	Are all pathways located outside of the equipment use zones or covered with synthetic impact-attenuating surfacing that meets the requirements of ASTM F 1292?
☐	☐	☐	15.	Is the restroom design safe for the neighborhood served?

TO CONTINUE AUDIT, COMPLETE ANNUAL OR PERIODIC INSPECTION.

©1997 MIG Communications

SITE SURVEY

PARK NAME _____ DATE OF INSPECTION _____ INSPECTOR _____

Yes	No	N/A	**ANNUAL OR PERIODIC INSPECTION**
☐	☐	☐	16. Is the play area readily visible from the street?
☐	☐	☐	17. Is the play area free of dangerous features, such as sudden drop-offs, drainage pipes, and ditches?
☐	☐	☐	18. Is the play area free of puddles and boggy, slow-draining soil conditions?
☐	☐	☐	19. Are all drainage grates locked?
☐	☐	☐	20. Do the openings to drainage grates run perpendicular to the path of travel to prevent hazards to bicyclists?
☐	☐	☐	21. Are all utility boxes locked?
☐	☐	☐	22. Is perimeter fencing in good condition, with no gap greater than 2 inches (50 mm) between the fence and the ground surface?
☐	☐	☐	23. Is the play area, pathways, and entrance free of trip hazards, such as protruding roots and abrupt changes in pavement level?
☐	☐	☐	24. Are hard-surfaced pathways free of slippery materials, such as loose-fill surfacing, sand, ice, snow, algae, or moss?
☐	☐	☐	25. Are all irrigation heads retracted?
☐	☐	☐	26. Are all sinks and drinking fountains clean, unclogged, and functioning?
☐	☐	☐	27. Are restrooms clean and functioning?
☐	☐	☐	28. Are all storm drains unclogged?
☐	☐	☐	29. Are play equipment platforms, stairs, or rungs free of slippery materials, such as water, ice, sand, or leaves?
☐	☐	☐	30. Is the play area free of loose, missing, or broken parts?
☐	☐	☐	31. Is the play area free of sharp edges?
☐	☐	☐	32. Has all scrap material such as ropes, sheets, and cardboard added to the play equipment been removed?
☐	☐	☐	33. Is only cold, potable water provided for the play area?
☐	☐	☐	34. Are trees, shrubs, groundcovers, and turf located outside of equipment use zones?

©1997 MIG Communications

SITE SURVEY

PARK NAME | DATE OF INSPECTION | INSPECTOR

Yes	No	N/A	**ANNUAL OR PERIODIC INSPECTION (cont.)**
☐	☐	☐	35. In areas where children may run through, are plants with flexible branches used?
☐	☐	☐	36. Do guying and staking methods meet protrusion and entrapment criteria?
☐	☐	☐	37. Is the play area free of diagonal guy wires that could cause a trip hazard?
☐	☐	☐	38. Is the play area free of plant materials that attract harmful pests?
☐	☐	☐	39. Is the play area free of undesirable plants, such as plants with thorns, plants that excrete sticky sap, trees that drop large limbs, and plants that require excessive pruning to discourage climbing?
☐	☐	☐	40. Is the play area free of trees and shrubs that shed plant parts excessively where shedding might create a hazard, such as over pathways?
☐	☐	☐	41. Is the play area free of mushrooms and other fungi?
☐	☐	☐	42. Are short turf species that are less likely to conceal sharp or foreign objects used in the play area?
☐	☐	☐	43. Is the play area free of vegetation that is poisonous when ingested or vegetation that may cause contact dermatitis?
☐	☐	☐	44. Are plants with berries or fruit avoided in play areas intended for children under 3 years?
☐	☐	☐	45. Are plant materials located to avoid creating visual barriers? ▷ *Clear sight-lines should be maintained from the street or nearby housing into the play area. In addition, all parts of the play area should be visible from more than one location within the play area.*
☐	☐	☐	46. Is the play area free of insect infestation?
☐	☐	☐	47. Is the play area free of rodents and other vertebrate pests?
☐	☐	☐	48. Is the use of pesticides and herbicides avoided in the play area?

©1997 MIG Communications

2 SAFETY SURFACING

_____ _____ _____
PARK NAME   DATE OF INSPECTION   INSPECTOR

Note: CPSC and ASTM do not address general site safety or specifications for loose-fill safety surfaces. The questions below addressing general safety considerations reflect the opinion of the authors. The questions below addressing material depth and specifications for loose-fill safety surfacing also reflect the opinion of the authors, and are based on tests prescribed by ASTM F 1292 and conducted by the authors at an independent testing laboratory.

Yes	No	N/A		
				Synthetic Surfacing: AUDIT
☐	☐	☐	1.	Is the surface guaranteed by the manufacturer to meet ASTM F 1292 standards for impact attenuation? ▷ *According to these standards, a head-first fall from the highest accessible height of the play equipment must not result in an impact of more than 200 g's or an HIC value of more than 1,000 (see Definitions).*
☐	☐	☐	2.	Have cutouts been filled with sealant to eliminate voids at equipment?

TO CONTINUE AUDIT, COMPLETE ANNUAL OR PERIODIC INSPECTION.

Yes	No	N/A		
				Synthetic Surfacing: ANNUAL OR PERIODIC INSPECTION
☐	☐	☐	3.	Are poured-in-place surfaces and synthetic tiles free of loose material and foreign objects, such as debris, sand, wood chips, gravel, leaves, soil, and toys?
☐	☐	☐	4.	Are poured-in-place surfaces and synthetic tiles free of puddles, ice, and snow?
☐	☐	☐	5.	Are poured-in-place surfaces and synthetic tiles firmly attached to the underlying surface?
☐	☐	☐	6.	Are poured-in-place surfaces and synthetic tiles free of abrupt changes in level greater than ¼ inch (6 mm)?
☐	☐	☐	7.	Are poured-in-place surfaces and synthetic tiles free of cuts, nicks, or damaged areas?
☐	☐	☐	8.	Are synthetic tiles free of exposed hardware and sharp edges?

©1997 MIG Communications

2 SAFETY SURFACING

PARK NAME DATE OF INSPECTION INSPECTOR

Yes	No	N/A	

Wood-Product Surfacing: AUDIT

☐ ☐ ☐   9. Are the wood-product areas free of asphalt or concrete subsurface materials as recommended by CPSC?
▷ *Wood products should be installed over a compacted subgrade, or, when manufactured as a safety surface, as recommended by the manufacturer.*

☐ ☐ ☐   10. Do the wood products meet ASTM F 1292 standards for impact attenuation?
▷ *According to these standards, a head-first fall from the highest accessible height of the play equipment must not result in an impact of more than 200 g's or an HIC value of more than 1,000 (see Definitions).*

TO CONTINUE AUDIT, COMPLETE ANNUAL AND PERIODIC INSPECTIONS.

Wood-Product Surfacing: ANNUAL INSPECTION

☐ ☐ ☐   11. Is the play equipment area closed when the wood-product surface is frozen?

☐ ☐ ☐   12. Are wood products removed and replaced when impact-attenuating ability deteriorates due to dirt, debris, or compaction?

TO CONTINUE ANNUAL INSPECTION, COMPLETE PERIODIC INSPECTION.

2 SAFETY SURFACING

| PARK NAME | DATE OF INSPECTION | INSPECTOR |

Wood-Product Surfacing: PERIODIC INSPECTION

Yes / No / N/A

- ☐ ☐ ☐ 13. Are the wood products free of debris and foreign objects, such as stones, leaves, twigs, branches, toys, broken glass, or other sharp objects?

- ☐ ☐ ☐ 14. Are the wood products free of animal feces?

- ☐ ☐ ☐ 15. Are the wood products free of mold, mushrooms, fungi, mildew, rot, and insect or rodent infestation?

- ☐ ☐ ☐ 16. Are the wood products contained in the surfacing area or removed from adjacent areas and pathways?

- ☐ ☐ ☐ 17. Are the wood products free of holes or low areas caused by digging or play activities?
  ▷ *Wood products require continuous maintenance to ensure a uniform depth and proper thickness for impact attenuation.*

- ☐ ☐ ☐ 18. Are the wood products free of puddles and poor drainage?

- ☐ ☐ ☐ 19. Are the wood products at least 12 inches (300 mm) deep throughout the use zone?

- ☐ ☐ ☐ 20. Do the wood products meet the following specifications for materials?

WOOD PRODUCT	MATERIAL SPECIFICATION
bark mulch	untreated chipped bark with a maximum size of 1 1/2 inches (40 mm) and no twigs, leaves, branches, thorns, dirt, or poisonous plants
wood mulch	untreated chipped tree prunings with a maximum size of 1 1/2 inches (40 mm) and no thorns, dirt, or poisonous plants
manufactured wood chips	particles varying in size from 1/8 to 1/2 inch (3 to 15 mm) thick by 1 to 3 inches (25 to 75 mm) long

©1997 MIG Communications

2 SAFETY SURFACING

PARK NAME   DATE OF INSPECTION   INSPECTOR

Yes   No   N/A

Sand Surfacing: AUDIT

☐ ☐ ☐   21. Are the sand areas free of asphalt or concrete subsurface materials as recommended by CPSC?
▷ *Sand should be installed over a compacted subgrade.*

☐ ☐ ☐   22. Does the sand meet ASTM F 1292 standards for impact attenuation?
▷ *According to these standards, a head-first fall from the highest accessible height of the play equipment must not result in an impact of more than 200 g's or an HIC value of more than 1,000 (see Definitions.).*

TO CONTINUE AUDIT, COMPLETE ANNUAL AND PERIODIC INSPECTIONS.

Sand Surfacing: ANNUAL INSPECTION

☐ ☐ ☐   23. Is the play equipment area closed when the sand surfacing is frozen?

☐ ☐ ☐   24. Has the sand been thoroughly sifted for cleaning and aeration on an annual basis?

☐ ☐ ☐   25. Is sand removed and replaced when impact-attenuating ability deteriorates due to dirt, debris, or compaction?

TO CONTINUE ANNUAL INSPECTION, COMPLETE PERIODIC INSPECTION.

2 SAFETY SURFACING

| PARK NAME | DATE OF INSPECTION | INSPECTOR |

Yes	No	N/A	

Sand Surfacing: PERIODIC INSPECTION

□ □ □ 26. Is the sand free of debris and foreign objects, such as stones, leaves, twigs, branches, toys, broken glass, or other sharp objects?

□ □ □ 27. Is the sand free of animal feces?

□ □ □ 28. Is the sand contained in the surfacing area or removed from adjacent areas and pathways?

□ □ □ 29. Is the sand free of holes or low areas caused by digging or play activities?
▷ *Sand requires continuous maintenance to ensure a uniform depth and proper thickness for impact attenuation.*

□ □ □ 30. Is the sand free of insect infestation?

□ □ □ 31. Is the sand free of puddles and poor drainage?
▷ *Sand is not recommended for use as a safety surface in wet climates because its impact-attenuating ability is greatly reduced when wet.*

□ □ □ 32. Is the sand at least 18 inches (450 mm) deep throughout the use zone?

□ □ □ 33. Is the sand rounded (by natural or mechanical means); washed; free of dust, clay, soil, hazardous substances, or foreign objects; and sieved as shown in the following table?

SIEVE SIZE	PERCENT PASSING
3/8 inch (10 mm)	100 percent
#4	99–100 percent
#8	81–95 percent
#16	53–75 percent
#30	35–56 percent
#50	20–25 percent
#100	5–9 percent
#200	less than 2 percent

2 SAFETY SURFACING

PARK NAME | DATE OF INSPECTION | INSPECTOR

Yes | No | N/A

Gravel Surfacing: AUDIT

☐ ☐ ☐ 34. Are the gravel areas free of asphalt or concrete subsurface materials as recommended by CPSC?
▷ *Gravel should be installed over a compacted subgrade.*

☐ ☐ ☐ 35. Does the gravel meet ASTM F 1292 standards for impact attenuation?
▷ *According to these standards, a head-first fall from the highest accessible height of the play equipment must not result in an impact of more than 200 g's or an HIC value of more than 1,000 (see Definitions).*

TO CONTINUE AUDIT, COMPLETE ANNUAL AND PERIODIC INSPECTIONS.

Gravel Surfacing: ANNUAL INSPECTION

☐ ☐ ☐ 36. Is the play equipment area closed when the gravel surfacing is frozen?

☐ ☐ ☐ 37. Has the gravel been thoroughly sifted for cleaning and aeration on an annual basis?

☐ ☐ ☐ 38. Is gravel removed and replaced when impact-attenuating ability deteriorates due to dirt, debris, or compaction?

TO CONTINUE ANNUAL INSPECTION, COMPLETE PERIODIC INSPECTION.

©1997 MIG Communications

2 SAFETY SURFACING

| PARK NAME | DATE OF INSPECTION | INSPECTOR |

Gravel Surfacing: PERIODIC INSPECTION

Yes | No | N/A

- [ ] [ ] [ ] 39. Is the gravel free of debris and foreign objects, such as stones, leaves, twigs, branches, toys, broken glass, or other sharp objects?

- [ ] [ ] [ ] 40. Is the gravel free of animal feces?

- [ ] [ ] [ ] 41. Is the gravel contained in the surfacing area or removed from adjacent areas and pathways?

- [ ] [ ] [ ] 42. Is the gravel free of holes or low areas caused by digging or play activities?
  ▷ *Gravel requires continuous maintenance to ensure a uniform depth and proper thickness for impact attenuation.*

- [ ] [ ] [ ] 43. Is the gravel free of insect infestation?

- [ ] [ ] [ ] 44. Is the gravel free of puddles and poor drainage?

- [ ] [ ] [ ] 45. Is the gravel at least 12 inches (300 mm) deep throughout the use zone?

- [ ] [ ] [ ] 46. Is the gravel rounded (by natural or mechanical means); washed; free of dust, clay, soil, hazardous substances, or foreign objects; and sieved as shown in the following table?

SIEVE SIZE	PERCENT PASSING
1/2 inch (15 mm)	100 percent
3/8 inch (10 mm)	75–85 percent

©1997 MIG Communications

2 SAFETY SURFACING

| PARK NAME | DATE OF INSPECTION | INSPECTOR |

Yes	No	N/A	

Chopped-Tire Surfacing: AUDIT

☐ ☐ ☐   47. Are the chopped-tire areas free of asphalt or concrete subsurface materials?
▷ *Chopped tire should be installed over a subgrade of compacted gravel covered with geotextile fabric.*

☐ ☐ ☐   48. Does the chopped tire meet ASTM F 1292 standards for impact attenuation?
▷ *According to these standards, a head-first fall from the highest accessible height of the play equipment must not result in an impact of more than 200 g's or an HIC value of more than 1,000 (see Definitions).*

TO CONTINUE AUDIT, COMPLETE ANNUAL AND PERIODIC INSPECTIONS.

Chopped-Tire Surfacing: ANNUAL INSPECTION

☐ ☐ ☐   49. Has the chopped tire been thoroughly sifted for cleaning on an annual basis?

TO CONTINUE ANNUAL INSPECTION, COMPLETE PERIODIC INSPECTION.

Chopped-Tire Surfacing: PERIODIC INSPECTION

☐ ☐ ☐   50. Is the chopped tire free of debris and foreign objects, such as stones, leaves, twigs, branches, toys, broken glass, or other sharp objects?

☐ ☐ ☐   51. Is the chopped tire free of animal feces?

☐ ☐ ☐   52. Is the chopped tire contained in the surfacing area or removed from adjacent areas and pathways?

☐ ☐ ☐   53. Is the chopped tire free of holes or low areas caused by digging or play activities?
▷ *Chopped tire requires continuous maintenance to ensure a uniform depth and proper thickness for impact attenuation.*

☐ ☐ ☐   54. Is the chopped tire free of insect infestation?

☐ ☐ ☐   55. Is the chopped tire free of puddles and poor drainage?

☐ ☐ ☐   56. Is the chopped-tire surfacing maintained at the depth recommended by the manufacturer?

☐ ☐ ☐   57. Is the chopped-tire surfacing free of metal, harmful chemicals, and foreign material?

©1997 MIG Communications

3 EQUIPMENT ACCESS & EGRESS

PARK NAME   DATE OF INSPECTION   INSPECTOR

rung ladder

minimum width
under 5 years: 12 in.
5 to 12 years: 16 in.

rung size
0.95 – 1.55 in.

maximum distance between rungs
2 to 12 years: 12 in.

stairway

maximum vertical rise
under 5 years: 9 in.
5 to 12 years: 12 in.

maximum slope
50°

Note: Questions relating to wheelchair-accessible ramps and transfer points are not based on ASTM or CPSC recommendations; these questions reflect the opinion of the authors. As of March 1997, the final ADA requirements for wheelchair-accessible ramps and transfer points were still under development. Contact the U.S. Architectural and Transportation Compliance Board at (202) 272-5434 for updated information and guidelines.

Yes   No   N/A   **AUDIT**

General Considerations

1. Are vertical angles greater than 55 degrees?
   ▷ *Inverted angles or angles with a filled apex are exempt (see Definitions).*

2. On rung ladders, net climbers, and arch climbers used for play equipment access, is the stepping surface used for final access located evenly with the play surface it serves?
   ▷ *Connecting play events above this point creates a potential head and neck entrapment and a trip hazard.*

3. Are steps, closed risers, ramps, and platforms designed so that they do not accumulate water, sand, or other debris?

4. Are ladder rungs and steps evenly spaced within a tolerance of 0.25 inch (6 mm) and horizontal within a tolerance of 2 degrees?

3 EQUIPMENT ACCESS & EGRESS

PARK NAME _____   DATE OF INSPECTION _____   INSPECTOR _____

ramp intended for wheelchair access

minimum clear width
36 in.

maximum slope
1:12

height of transfer platform
2 to 5 years: 11 to 14 in.
5 to 12 years: 14 to 17 in.

transfer point

Yes	No	N/A	
			AUDIT (cont.)
☐	☐	☐	5. Are platforms and play surfaces horizontal within a tolerance of 2 degrees?
☐	☐	☐	6. For 2- to 5-year-olds, are ramps or stairways with closed risers provided? ▷ *For 2- to 5-year-olds, rung ladders and step ladders may be provided if a less difficult means of access and egress is also included.*
			Rung Ladders
☐	☐	☐	7. a. For 2- to 5-year-olds, are rung ladders at least 12 inches (300 mm) wide?
☐	☐	☐	b. For 5- to 12-year-olds, are rung ladders at least 16 inches (400 mm) wide?
☐	☐	☐	8. Do rung ladders have a slope of 75 to 90 degrees?
☐	☐	☐	9. For 2- to 12-year-olds, is the distance between rungs (tread-to-tread vertical rise) no more than 12 inches (300 mm)?
☐	☐	☐	10. Do the rungs measure between 0.95 and 1.55 inches (24.1 and 39.4 mm) in diameter?

©1997 MIG Communications

3 EQUIPMENT ACCESS & EGRESS

PARK NAME DATE OF INSPECTION INSPECTOR

Yes   No   N/A   **AUDIT (cont.)**

Stepladders

☐ ☐ ☐ 11. Do stepladders have a slope of 50 to 75 degrees?

☐ ☐ ☐ 12. a. For 2- to 5-year-olds, do stepladders for single-file use have a tread width between 12 and 21 inches (300 and 530 mm)?
▷ *Stepladders designed for use by two children abreast are not recommended for this age group.*

☐ ☐ ☐ b. For 5- to 12-year-olds, do stepladders for single-file use have a minimum tread width of 16 inches (400 mm)?

☐ ☐ ☐ c. For 5- to 12-year-olds, do stepladders for use by two children abreast have a minimum tread width of 36 inches (910 mm)?

☐ ☐ ☐ 13. a. For 2- to 5-year-olds, do stepladders with open or closed risers have a minimum tread depth of 7 inches (180 mm)?

☐ ☐ ☐ b. For 5- to 12-year-olds, do stepladders with open risers have minimum tread depth of 3 inches (76 mm)?

☐ ☐ ☐ c. For 5- to 12-year-olds, do stepladders with closed risers have minimum tread depth of 6 inches (150 mm)?

☐ ☐ ☐ 14. a. For 2- to 5-year-olds, is the distance between stepladder rungs (tread-to-tread vertical rise) no more than 9 inches (228 mm)?

☐ ☐ ☐ b. For 5- to 12-year-olds, is the distance between stepladder rungs (tread-to-tread vertical rise) no more than 12 inches (300 mm)?

Stairways

☐ ☐ ☐ 15. Do stairways have a maximum slope of 50 degrees?

☐ ☐ ☐ 16. a. For 2- to 5-year-olds, do stairways for single-file use have a minimum tread width of 12 inches (300 mm)?

☐ ☐ ☐ b. For 2- to 5-year-olds, do stairways for use by two children abreast have a minimum tread width of 30 inches (760 mm)?

☐ ☐ ☐ c. For 5- to 12-year-olds, do stairways for single-file use have a minimum tread width of 16 inches (400 mm)?

☐ ☐ ☐ d. For 5- to 12-year-olds, do stairways for use by two children abreast have a minimum tread width of 36 inches (910 mm)?

©1997 MIG Communications

3 EQUIPMENT ACCESS & EGRESS

RECOMMENDED DIMENSIONS FOR LADDERS, STAIRWAYS, AND RAMPS

rung ladders	2 to 5 years	5 to 12 years
slope	75–90 degrees	75–90 degrees
rung width	≥ 12 in. (300 mm)	≥ 16 in. (400 mm)
vertical rise (tread to tread)	≤ 12 in. (300 mm)	≤ 12 in. (300 mm)
rung diameter	0.95–1.55 in. (24.1–39.4 mm)	0.95–1.55 in. (24.1–39.4 mm)

stepladders	2 to 5 years	5 to 12 years
slope	50–75 degrees	50–75 degrees
tread width		
single-file access	12–21 in. (300–530 mm)	≥ 16 in. (400 mm)
two-abreast access	n/a	≥ 36 in. (910 mm)
tread depth		
open riser	≥ 7 in. (180 mm)	≥ 3 in. (76 mm)
closed riser	≥ 7 in. (180 mm)	≥ 6 in. (150 mm)
vertical rise (tread to tread)	≤ 9 in. (228 mm)	≤ 12 in. (300 mm)

stairways	2 to 5 years	5 to 12 years
slope	≤ 50 degrees	≤ 50 degrees
tread width		
single-file access	≥ 12 in. (300 mm)	≥ 16 in. (400 mm)
two-abreast access	≥ 30 in. (760 mm)	≥ 36 in. (910 mm)
tread depth		
open riser	n/a	≥ 8 in. (200 mm)
closed riser	≥ 7 in. (180 mm)	≥ 8 in. (200 mm)
vertical rise (tread to tread)	≤ 9 in. (228 mm)	≤ 12 in. (300 mm)

ramps*	2 to 5 years	5 to 12 years
slope (vertical:horizontal)	≤ 1:8	≤ 1:8
width		
single-file access	≥ 12 in. (300 mm)	≥ 16 in. (400 mm)
two-abreast access	≥ 30 in. (760 mm)	≥ 36 in. (910 mm)

*not intended for wheelchair use

wheelchair-accessible ramps	2 to 5 years	5 to 12 years
slope (vertical:horizontal)	≤ 1:12	≤ 1:12
clear width	≥ 36 in. (910 mm)	≥ 36 in. (910 mm)
cross slope	≤ 1:50	≤ 1:50
horizontal run	≤ 144 in. (3700 mm)	≤ 144 in. (3700 mm)

©1997 MIG Communications

3 EQUIPMENT ACCESS & EGRESS

| PARK NAME | DATE OF INSPECTION | INSPECTOR |

Yes   No   N/A

AUDIT (cont.)

☐ ☐ ☐ 17. a. For 2- to 5-year-olds, do stairways with closed risers have a minimum tread depth of 7 inches (180 mm)?
▷ *Stairways with open risers are not recommended for this age group.*

☐ ☐ ☐ b. For 5- to 12-year-olds, do stairways with either open or closed risers have a minimum tread depth of 8 inches (200 mm)?

☐ ☐ ☐ c. For 5- to 12-year-olds, do spiral stairways have a minimum tread depth of 8 inches (200 mm) at the outer edge of the steps?

☐ ☐ ☐ 18. a. For 2- to 5-year-olds, is the distance between steps (tread-to-tread vertical rise) no more than 9 inches (228 mm)?

☐ ☐ ☐ b. For 5- to 12-year-olds, is the distance between steps (tread-to-tread vertical rise) no more than 12 inches (300 mm)?

Ramps (Not Intended for Wheelchair Access)

☐ ☐ ☐ 19. Do ramps not intended for wheelchair access have a maximum slope of 1:8?

☐ ☐ ☐ 20. a. For 2- to 5-year-olds, do ramps (not intended for wheelchair access) for single-file use have a minimum width of 12 inches (300 mm)?

☐ ☐ ☐ b. For 2- to 5-year-olds, do ramps for use by two children abreast have a minimum width of 30 inches (760 mm)?

☐ ☐ ☐ c. For 5- to-12 year-olds, do ramps for single-file use have a minimum width of 16 inches (400 mm)?

☐ ☐ ☐ d. For 5- to-12 year-olds, do ramps for use by two children abreast have a minimum width of 36 inches (910 mm)?

Wheelchair-Accessible Ramps

☐ ☐ ☐ 21. Do accessible ramps (i.e., intended for wheelchair access) have a minimum clear width of 36 inches (910 mm)?

☐ ☐ ☐ 22. Do accessible ramps have a maximum slope of 1:12?

☐ ☐ ☐ 23. Do accessible ramps have a maximum cross slope of 1:50?

☐ ☐ ☐ 24. Do accessible ramps have a horizontal run no greater than 144 inches (3700 mm)?

©1997 MIG Communications

3 EQUIPMENT ACCESS & EGRESS

PARK NAME _____ DATE OF INSPECTION _____ INSPECTOR _____

Yes	No	N/A	

AUDIT (cont.)

☐ ☐ ☐ 25. a. Are landings provided at the bottom and top of each accessible ramp?

☐ ☐ ☐   b. Are the landings at least as wide as the ramps leading to them?

☐ ☐ ☐   c. On landings that include play events, is there a space where a wheelchair user can park and play?
       ▷ *The parking space should measure at least 30 by 48 inches (760 by 1200 mm). It should not reduce the width of the circulation path to less than 36 inches (900 mm).*

☐ ☐ ☐ 26. Do the landings measure at least 60 by 60 inches (1525 by 1525 mm)?

☐ ☐ ☐ 27. Is an appropriate edge treatment provided to keep wheelchairs from rolling off ramps and landings?

☐ ☐ ☐ 28. Do accessible ramps have a slip-resistant, water-resistant surface?

☐ ☐ ☐ 29. Where the ramp connects with the ground and adjoining landing, is the connection point flush with the adjoining surface?

Handrails and Access Devices

☐ ☐ ☐ 30. Do stairways and stepladders have continuous handrails on both sides?
       ▷ *Rung ladders do not require handrails since the rungs or side supports serve this function. Spiral stairways that do not permit handrails on both sides should have a continuous handrail along the outside perimeter of the steps.*

☐ ☐ ☐ 31. Is the height of the handrail for stairways and stepladders between 22 and 38 inches (560 and 970 mm), measured from the top front edge of the step to the top surface of the handrail?

☐ ☐ ☐ 32. Do handrails measure between 0.95 and 1.55 inches (24.1 and 39.4 mm) in diameter?

☐ ☐ ☐ 33. On access devices or play events that do not use handrails, are handgrips provided to facilitate the transition to a platform?
       ▷ *Rung ladders, arch climbers, and flexible climbers should have handgrips.*

☐ ☐ ☐ 34. Do handrails have rounded ends or ends that return to the wall, ground, platform, or post?

©1997 MIG Communications

3 EQUIPMENT ACCESS & EGRESS

PARK NAME   DATE OF INSPECTION   INSPECTOR

Yes   No   N/A

AUDIT (cont.)

☐ ☐ ☐   35. a. For 2- to 5-year-olds, do accessible ramps have handrails between 12 and 16 inches (305 and 410 mm) high along both sides of the ramp?

☐ ☐ ☐   b. For 5- to 12-year-olds, do accessible ramps have handrails between 20 and 28 inches (500 and 710 mm) high along both sides of the ramp?

Stepped Platforms

☐ ☐ ☐   36. a. For 2- to 5-year-olds, do stepped platforms have a maximum height difference of 12 inches (300 mm)?

☐ ☐ ☐   b. For 5- to 12-year-olds, do stepped platforms have a maximum height difference of 18 inches (460 mm)?

Transfer Points

☐ ☐ ☐   37. a. For 2- to 5-year-olds, are transfer points located at a height of 11 to 14 inches (275 to 350 mm) above the accessible route or platform?

☐ ☐ ☐   b. For 5- to 12-year-olds, are transfer points located at a height of 14 to 17 inches (350 to 425 mm) above the accessible route or platform?

☐ ☐ ☐   38. Are transfer points at least 24 inches (610 mm) wide?

☐ ☐ ☐   39. Are transfer points at least 14 inches (360 mm) deep?
▷ *Further research is needed to verify whether or not this depth is adequate. A depth greater than 14 inches may be required.*

☐ ☐ ☐   40. Do transfer points have handrails to assist wheelchair users?

☐ ☐ ☐   41. Do steps and platforms adjacent to transfer points have closed risers to prevent potential entrapment?

☐ ☐ ☐   42. a. For 2- to 5-year-olds, do steps adjacent to transfer points have a maximum step height of 6 inches (150 mm)?

☐ ☐ ☐   b. For 5- to 12-year-olds, do steps adjacent to transfer points have a maximum step height of 8 inches (200 mm)?

☐ ☐ ☐   43. Is a wheelchair turning space measuring at least 60 inches (1525 mm) in diameter provided at the base of transfer points?

©1997 MIG Communications

3 EQUIPMENT ACCESS & EGRESS

PARK NAME **DATE OF INSPECTION** **INSPECTOR**

Yes No N/A

AUDIT (cont.)

☐ ☐ ☐ 44. Is there a parking area measuring at least 30 by 48 inches (760 by 1200 mm) adjacent to the transfer platform off the accessible route of travel?
▷ *The parking area should not overlap with or reduce the width of the circulation path to less than 36 inches (910 mm).*

TO CONTINUE AUDIT, COMPLETE APPROPRIATE EQUIPMENT CHECKLISTS.

4 GUARDRAILS & PROTECTIVE BARRIERS

PARK NAME   DATE OF INSPECTION   INSPECTOR

guardrail

maximum height of lower edge
2 to 5 years: 23 in.
5 to 12 years: 28 in.

minimum height of top edge
2 to 5 years: 29 in.
5 to 12 years: 38 in.

protective barrier

top of protective barrier

minimum enclosure height
2 to 5 years: 29 in.
5 to 12 years: 38 in.

Yes	No	N/A	**AUDIT**
☐	☐	☐	1. Does the equipment meet all standards for structural integrity as specified by ASTM F 1487?
☐	☐	☐	2. Are vertical angles greater than 55 degrees? ▷ *Inverted angles or angles with a filled apex are exempt (see Definitions).*
☐	☐	☐	3. a. For 2- to 5-year-olds, are all play equipment platforms over 30 inches (760 mm) high enclosed by a protective barrier at least 29 inches (740 mm) high?
☐	☐	☐	b. For 5- to 12-year-olds, are all play equipment platforms over 48 inches (1200 mm) high enclosed by a protective barrier at least 38 inches (970 mm) high? ▷ *Game panels that meet design criteria for protective barriers are acceptable.*
☐	☐	☐	4. a. For 2- to 5-year-olds, are all play equipment platforms over 20 inches (510 mm) high enclosed by a guardrail that is a maximum 23 inches (580 mm) high at the lower edge and 29 inches (740 mm) high at the top edge?
☐	☐	☐	b. For 5- to 12-year-olds, are all play equipment platforms over 30 inches (760 mm) high enclosed by a guardrail that is a maximum 28 inches (710 mm) high at the lower edge and 38 inches (970 mm) high at the top edge?

©1997 MIG Communications

4 GUARDRAILS & PROTECTIVE BARRIERS

PARK NAME _____ DATE OF INSPECTION _____ INSPECTOR _____

Yes	No	N/A	AUDIT (cont.)
☐	☐	☐	5. Is the equipment free of extra holes that could harbor nesting insects? ▷ *This question is based on the authors' opinion and is not addressed by CPSC or ASTM.*
☐	☐	☐	6. Is the equipment free of pinch, crush, and shear points (see Definitions)?
☐	☐	☐	7. Is the equipment free of cables, wires, or other suspended hazards hung within 45 degrees of horizontal (see Definitions)?
☐	☐	☐	8. Is the equipment free of handholds or footholds that may facilitate climbing?
☐	☐	☐	9. In areas between adjacent platforms that do not permit a protective barrier of the recommended height, is protective infill provided?
☐	☐	☐	10. Are lock washers, self-locking nuts, or other locking means provided for all nuts and bolts to protect them from detachment?
☐	☐	☐	11. Do all metal edges have rolled edging or rounded capping?
☐	☐	☐	12. Are metal materials painted, galvanized, anodized, or composed of non-rusting materials?
☐	☐	☐	13. When located in direct sunlight, have metal materials been coated in plastic to avoid the risk of a contact-burn injury? ▷ *Bare or painted metal surfaces should be avoided in intense, direct sunlight.*
☐	☐	☐	14. Are plastic materials ultraviolet-stabilized to resist fading? ▷ *This question is based on the authors' opinion and is not addressed by CPSC or ASTM.*

TO CONTINUE AUDIT, COMPLETE ANNUAL AND PERIODIC INSPECTIONS.

4 GUARDRAILS & PROTECTIVE BARRIERS

PARK NAME   DATE OF INSPECTION   INSPECTOR

Yes	No	N/A	**ANNUAL INSPECTION**
☐	☐	☐	15. Is the equipment free of head and neck entrapments (see Inspection Procedures)? ▷ *For protective barriers and game panels functioning as barriers, openings within the barrier and between the barrier's lower edge and platform surface should preclude the passage of the torso probe.*
☐	☐	☐	16. Do protrusions meet the protrusion test criteria (see Inspection Procedures)?
☐	☐	☐	17. Is the equipment free of hollow support posts or tubes with open ends?
☐	☐	☐	18. Are equipment footings securely anchored?
☐	☐	☐	19. Are wood materials naturally rot- and insect-resistant, or treated with a wood preservative below and up to 6 inches (150 mm) above the surface of the play area? If a wood preservative was used, list the preservative's name: _____.
☐	☐	☐	20. Is the wood preservative safe for use in children's play areas, as specified by ASTM F 1487 standards?
☐	☐	☐	21. Are paints free of lead (0.06% maximum lead by dry weight) as specified by ASTM F 1487 standards?

TO CONTINUE ANNUAL INSPECTION, COMPLETE PERIODIC INSPECTION.

4 GUARDRAILS & PROTECTIVE BARRIERS

PARK NAME   DATE OF INSPECTION   INSPECTOR

Yes	No	N/A	**PERIODIC INSPECTION**
☐	☐	☐	22. Is the equipment stable and without severe structural deterioration, such as at the footings and joints?
☐	☐	☐	23. Is the equipment free of loose, missing, or broken parts and vandalism?
☐	☐	☐	24. Is the equipment free of sharp points, corners, or edges?
☐	☐	☐	25. Is all hardware present, securely attached, and free of significant wear? ▷ *Wear is indicated by visible elongation, deformation, indentation, rust, corrosion, or stripping.*
☐	☐	☐	26. Do bolt ends extend no more than two threads beyond the face of the nut?
☐	☐	☐	27. Are all fastening devices closed to prevent entanglement (see Definitions)?
☐	☐	☐	28. Are wood materials free of warping, wood rot, insect damage, cupping, and checking?
☐	☐	☐	29. Are wood materials free of splinters, heart center, and loose or missing knots?
☐	☐	☐	30. Are metal materials free of rust, corrosion, peeling paint, and bent parts?
☐	☐	☐	31. Are plastic parts unbroken, unchipped, and uncracked, particularly at joints and connections?
☐	☐	☐	32. Is the equipment free of chipped, peeling, or worn paint?

©1997 MIG Communications

5 BALANCE BEAMS

_____ _____ _____
PARK NAME   DATE OF INSPECTION   INSPECTOR

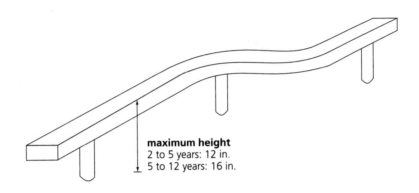

maximum height
2 to 5 years: 12 in.
5 to 12 years: 16 in.

Yes	No	N/A	AUDIT
☐	☐	☐	1. Does the balance beam meet all standards for structural integrity as specified by ASTM F 1487?
☐	☐	☐	2. Does the balance beam have a 72-inch (1800 mm) unobstructed use zone? ▷ *Two balance beams may have overlapping use zones.*
☐	☐	☐	3. Are vertical angles greater than 55 degrees? ▷ *Inverted angles or angles with a filled apex are exempt (see Definitions).*
☐	☐	☐	4. a. For 2- to 5-year-olds, is the balance beam no more than 12 inches (300 mm) high?
☐	☐	☐	b. For 5- to 12-year-olds, is the balance beam no more than 16 inches (410 mm) high?
☐	☐	☐	5. Is the balance beam free of extra holes that could harbor nesting insects? ▷ *This question is based on the authors' opinion and is not addressed by CPSC or ASTM.*
☐	☐	☐	6. Is the balance beam free of pinch, crush, and shear points (see Definitions)?

©1997 MIG Communications

5 BALANCE BEAMS

PARK NAME _____ DATE OF INSPECTION _____ INSPECTOR _____

Yes	No	N/A	AUDIT (cont.)
☐	☐	☐	7. Is the walking surface free of tripping hazards, such as beam support posts that extend above the level of the walking surface?
☐	☐	☐	8. Do chains or cables meet ASTM F 1487 structural integrity requirements?
☐	☐	☐	9. Do cables measure at least 1 inch (25 mm) in diameter?
☐	☐	☐	10. Are lock washers, self-locking nuts, or other locking means provided for all nuts and bolts to protect them from detachment?
☐	☐	☐	11. Do all metal edges have rolled edging or rounded capping?
☐	☐	☐	12. Are metal materials painted, galvanized, anodized, or composed of non-rusting materials?
☐	☐	☐	13. When located in direct sunlight, have metal materials been coated in plastic to avoid the risk of a contact-burn injury? ▷ *Bare or painted metal surfaces should be avoided in intense, direct sunlight.*
☐	☐	☐	14. Are plastic materials ultraviolet-stabilized to resist fading? ▷ *This question is based on the authors' opinion and is not addressed by CPSC or ASTM.*

TO CONTINUE AUDIT, COMPLETE ANNUAL AND PERIODIC INSPECTIONS.

5 BALANCE BEAMS

PARK NAME　　　　DATE OF INSPECTION　　　　INSPECTOR

Yes	No	N/A	**ANNUAL INSPECTION**
☐	☐	☐	15. Is the balance beam free of head and neck entrapments (see Inspection Procedures)?
☐	☐	☐	16. Do protrusions meet the protrusion test criteria (see Inspection Procedures)?
☐	☐	☐	17. Is the balance beam free of hollow support posts or tubes with open ends?
☐	☐	☐	18. Are equipment footings securely anchored?
☐	☐	☐	19. Are wood materials naturally rot- and insect-resistant, or treated with a wood preservative below and up to 6 inches (150 mm) above the surface of the play area?

If a wood preservative was used, list the preservative's name:

_____.

Yes	No	N/A	
☐	☐	☐	20. Is the wood preservative safe for use in children's play areas, as specified by ASTM F 1487 standards?
☐	☐	☐	21. Are paints free of lead (0.06% maximum lead by dry weight) as specified by ASTM F 1487 standards?

TO CONTINUE ANNUAL INSPECTION, COMPLETE PERIODIC INSPECTION.

5 BALANCE BEAMS

| PARK NAME | DATE OF INSPECTION | INSPECTOR |

Yes	No	N/A	**PERIODIC INSPECTION**
☐	☐	☐	22. Is the balance beam stable and without severe structural deterioration, such as at the footings and joints?
☐	☐	☐	23. Is the balance beam free of loose, missing, or broken parts and vandalism?
☐	☐	☐	24. Is the balance beam free of sharp points, corners, or edges?
☐	☐	☐	25. Are chains or cables without significant wear? ▷ *Wear is indicated by visible elongation, deformation, indentation, rust, or corrosion.*
☐	☐	☐	26. Are cables free of frayed or projecting wires?
☐	☐	☐	27. Are cables or chains fixed tightly at both ends so that there is no possibility of overlapping and entrapping a child?
☐	☐	☐	28. Is all hardware present, securely attached, and free of significant wear? ▷ *Wear is indicated by visible elongation, deformation, indentation, rust, corrosion, or stripping.*
☐	☐	☐	29. Do bolt ends extend no more than two threads beyond the face of the nut?
☐	☐	☐	30. Are all fastening devices closed to prevent entanglement (see Definitions)?
☐	☐	☐	31. Are wood materials free of warping, wood rot, insect damage, cupping, and checking?
☐	☐	☐	32. Are wood materials free of splinters, heart center, and loose or missing knots?
☐	☐	☐	33. Are metal materials free of rust, corrosion, peeling paint, and bent parts?
☐	☐	☐	34. Are plastic parts unbroken, unchipped, and uncracked, particularly at joints and connections?
☐	☐	☐	35. Is the balance beam free of chipped, peeling, or worn paint?

©1997 MIG Communications

6 BARS, CHIN-UP & TURNING

1 OF 4

PARK NAME DATE OF INSPECTION INSPECTOR

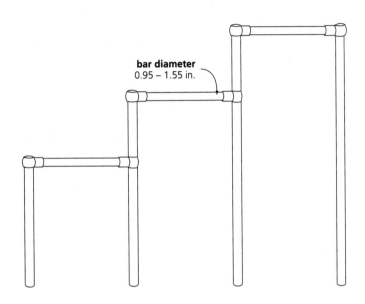

Yes	No	N/A	AUDIT
☐	☐	☐	1. Do the bars meet all standards for structural integrity as specified by ASTM F 1487?
☐	☐	☐	2. Do the bars have a 72-inch (1800 mm) unobstructed use zone?
☐	☐	☐	3. Are vertical angles greater than 55 degrees? ▷ *Inverted angles or angles with a filled apex are exempt (see Definitions).*
☐	☐	☐	4. Are the bars free of extra holes that could harbor nesting insects? ▷ *This question is based on the authors' opinion and is not addressed by CPSC or ASTM.*
☐	☐	☐	5. Are the bars free of pinch, crush, and shear points (see Definitions)?
☐	☐	☐	6. Are the bars free of cables, wires, or other suspended hazards hung within 45 degrees of horizontal (see Definitions)?
☐	☐	☐	7. Do the bars measure between 0.95 and 1.55 inches (24.1 and 39.4 mm) in diameter?

©1997 MIG Communications

6 BARS, CHIN-UP & TURNING

| PARK NAME | DATE OF INSPECTION | INSPECTOR |

Yes	No	N/A	**AUDIT (cont.)**
☐	☐	☐	8. Are lock washers, self-locking nuts, or other locking means provided for all nuts and bolts to protect them from detachment?
☐	☐	☐	9. Do all metal edges have rolled edging or rounded capping?
☐	☐	☐	10. Are metal materials painted, galvanized, anodized, or composed of non-rusting materials?
☐	☐	☐	11. When located in direct sunlight, have metal materials been coated in plastic to avoid the risk of a contact-burn injury? ▷ *Bare or painted metal surfaces should be avoided in intense, direct sunlight.*
☐	☐	☐	12. Are plastic materials ultraviolet-stabilized to resist fading? ▷ *This question is based on the authors' opinion and is not addressed by CPSC or ASTM.*

TO CONTINUE AUDIT, COMPLETE ANNUAL AND PERIODIC INSPECTIONS.

6 BARS, CHIN-UP & TURNING

| PARK NAME | DATE OF INSPECTION | INSPECTOR |

Yes	No	N/A	**ANNUAL INSPECTION**
☐	☐	☐	13. Are the bars free of head and neck entrapments (see Inspection Procedures)?
☐	☐	☐	14. Do protrusions meet the protrusion test criteria (see Inspection Procedures)?
☐	☐	☐	15. Are the bars free of hollow support posts or tubes with open ends?
☐	☐	☐	16. Are equipment footings securely anchored?
☐	☐	☐	17. Are wood materials naturally rot- and insect-resistant, or treated with a wood preservative below and up to 6 inches (150 mm) above the surface of the play area?

If a wood preservative was used, list the preservative's name:

_____.

Yes	No	N/A	
☐	☐	☐	18. Is the wood preservative safe for use in children's play areas, as specified by ASTM F 1487 standards?
☐	☐	☐	19. Are paints free of lead (0.06% maximum lead by dry weight) as specified by ASTM F 1487 standards?

TO CONTINUE ANNUAL INSPECTION, COMPLETE PERIODIC INSPECTION.

6 BARS, CHIN-UP & TURNING

| PARK NAME | DATE OF INSPECTION | INSPECTOR |

Yes No N/A

PERIODIC INSPECTION

☐ ☐ ☐   20. Are the bars stable and without severe structural deterioration, such as at the footings and joints?

☐ ☐ ☐   21. Are the bars free of loose, missing, or broken parts and vandalism?

☐ ☐ ☐   22. Are the bars free of sharp points, corners, or edges?

☐ ☐ ☐   23. Is all hardware present, securely attached, and free of significant wear?
            ▷ *Wear is indicated by visible elongation, deformation, indentation, rust, corrosion, or stripping.*

☐ ☐ ☐   24. Do bolt ends extend no more than two threads beyond the face of the nut?

☐ ☐ ☐   25. Are all fastening devices closed to prevent entanglement (see Definitions)?

☐ ☐ ☐   26. Are wood materials free of warping, wood rot, insect damage, cupping, and checking?

☐ ☐ ☐   27. Are wood materials free of splinters, heart center, and loose or missing knots?

☐ ☐ ☐   28. Are metal materials free of rust, corrosion, peeling paint, and bent parts?

☐ ☐ ☐   29. Are plastic parts unbroken, unchipped, and uncracked, particularly at joints and connections?

☐ ☐ ☐   30. Are the bars free of chipped, peeling, or worn paint?

©1997 MIG Communications

7 BARS, PARALLEL

PARK NAME **DATE OF INSPECTION** **INSPECTOR**

Note: According to ASTM F 1487, upper-body equipment requiring full support of body weight is not recommended for children under 5 years.

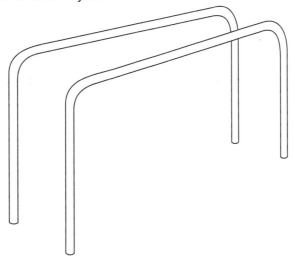

Yes	No	N/A	AUDIT
☐	☐	☐	1. Do the parallel bars meet all standards for structural integrity as specified by ASTM F 1487?
☐	☐	☐	2. Do the parallel bars have a 72-inch (1800 mm) unobstructed use zone?
☐	☐	☐	3. Are vertical angles greater than 55 degrees? ▷ *Inverted angles or angles with a filled apex are exempt (see Definitions).*
☐	☐	☐	4. Are the parallel bars free of extra holes that could harbor nesting insects? ▷ *This question is based on the authors' opinion and is not addressed by CPSC or ASTM.*
☐	☐	☐	5. Are the parallel bars free of pinch, crush, and shear points (see Definitions)?
☐	☐	☐	6. Are the parallel bars free of cables, wires, or other suspended hazards hung within 45 degrees of horizontal (see Definitions)?
☐	☐	☐	7. Are lock washers, self-locking nuts, or other locking means provided for all nuts and bolts to protect them from detachment?

©1997 MIG Communications

7 BARS, PARALLEL

PARK NAME _____ DATE OF INSPECTION _____ INSPECTOR _____

Yes	No	N/A	**AUDIT (cont.)**
☐	☐	☐	8. Do all metal edges have rolled edging or rounded capping?
☐	☐	☐	9. Are metal materials painted, galvanized, anodized, or composed of non-rusting materials?
☐	☐	☐	10. When located in direct sunlight, have metal materials been coated in plastic to avoid the risk of a contact-burn injury? ▷ *Bare or painted metal surfaces should be avoided in intense, direct sunlight.*
☐	☐	☐	11. Are plastic materials ultraviolet-stabilized to resist fading? ▷ *This question is based on the authors' opinion and is not addressed by CPSC or ASTM.*

TO CONTINUE AUDIT, COMPLETE ANNUAL AND PERIODIC INSPECTIONS.

7 BARS, PARALLEL

PARK NAME │ DATE OF INSPECTION │ INSPECTOR

Yes	No	N/A	**ANNUAL INSPECTION**
☐	☐	☐	12. Are the parallel bars free of head and neck entrapments (see Inspection Procedures)?
☐	☐	☐	13. Do protrusions meet the protrusion test criteria (see Inspection Procedures)?
☐	☐	☐	14. Are the parallel bars free of hollow support posts or tubes with open ends?
☐	☐	☐	15. Are equipment footings securely anchored?
☐	☐	☐	16. Are wood materials naturally rot- and insect-resistant, or treated with a wood preservative below and up to 6 inches (150 mm) above the surface of the play area?

If a wood preservative was used, list the preservative's name:

_____.

Yes	No	N/A	
☐	☐	☐	17. Is the wood preservative safe for use in children's play areas, as specified by ASTM F 1487 standards?
☐	☐	☐	18. Are paints free of lead (0.06% maximum lead by dry weight) as specified by ASTM F 1487 standards?

TO CONTINUE ANNUAL INSPECTION, COMPLETE PERIODIC INSPECTION.

©1997 MIG Communications

7 BARS, PARALLEL

| PARK NAME | DATE OF INSPECTION | INSPECTOR |

| Yes | No | N/A |

PERIODIC INSPECTION

- [ ] [ ] [ ] 19. Are the parallel bars stable and without severe structural deterioration, such as at the footings and joints?

- [ ] [ ] [ ] 20. Are the parallel bars free of loose, missing, or broken parts and vandalism?

- [ ] [ ] [ ] 21. Are the parallel bars free of sharp points, corners, or edges?

- [ ] [ ] [ ] 22. Is all hardware present, securely attached, and free of significant wear?
  ▷ *Wear is indicated by visible elongation, deformation, indentation, rust, corrosion, or stripping.*

- [ ] [ ] [ ] 23. Do bolt ends extend no more than two threads beyond the face of the nut?

- [ ] [ ] [ ] 24. Are all fastening devices closed to prevent entanglement (see Definitions)?

- [ ] [ ] [ ] 25. Are wood materials free of warping, wood rot, insect damage, cupping, and checking?

- [ ] [ ] [ ] 26. Are wood materials free of splinters, heart center, and loose or missing knots?

- [ ] [ ] [ ] 27. Are metal materials free of rust, corrosion, peeling paint, and bent parts?

- [ ] [ ] [ ] 28. Are plastic parts unbroken, unchipped, and uncracked, particularly at joints and connections?

- [ ] [ ] [ ] 29. Are the parallel bars free of chipped, peeling, or worn paint?

©1997 MIG Communications

8 BRIDGES, CLATTER

PARK NAME **DATE OF INSPECTION** **INSPECTOR**

Note: In the authors' opinion, clatter bridges are not recommended for children under 3.

maximum height
3 to 5 years: 30 in.
5 to 12 years: 48 in.

Yes	No	N/A	**AUDIT**
☐	☐	☐	1. Does the bridge meet all standards for structural integrity as specified by ASTM F 1487?
☐	☐	☐	2. Does the bridge have a 72-inch (1800 mm) unobstructed use zone?
☐	☐	☐	3. Are vertical angles greater than 55 degrees? ▷ *Inverted angles or angles with a filled apex are exempt (see Definitions).*
☐	☐	☐	4. a. For 3- to 5-year-olds, is the bridge surface no more than 30 inches (760 mm) high?
☐	☐	☐	b. For 5- to 12-year-olds, is the bridge surface no more than 48 inches (1200 mm) high? ▷ *The specified maximum height of the bridge surface allows the use of a guardrail, which is necessary for the function of the bridge.*
☐	☐	☐	5. a. Are guardrails provided to help prevent children from falling off the bridge?
☐	☐	☐	b. For 3- to 5-year-olds, is the top edge of the guardrail at least 29 inches (740 mm) high and the lower edge no more than 23 inches (580 mm) above the bridge walking surface?

8 BRIDGES, CLATTER

PARK NAME _____ DATE OF INSPECTION _____ INSPECTOR _____

Yes	No	N/A	AUDIT (cont.)
☐	☐	☐	5. c. For 5- to 12-year-olds, is the top edge of the guardrail at least 38 inches (970 mm) high and the lower edge no more than 28 inches (710 mm) above the bridge walking surface?
☐	☐	☐	6. Is the bridge free of extra holes that could harbor nesting insects? ▷ *This question is based on the authors' opinion and is not addressed by CPSC or ASTM.*
☐	☐	☐	7. Is the bridge free of pinch, crush, and shear points (see Definitions)?
☐	☐	☐	8. Is the bridge free of cables, wires, or other suspended hazards hung within 45 degrees of horizontal (see Definitions)?
☐	☐	☐	9. Do chains or cables meet ASTM F 1487 structural integrity requirements?
☐	☐	☐	10. Do cables measure at least 1 inch (25 mm) in diameter?
☐	☐	☐	11. Are lock washers, self-locking nuts, or other locking means provided for all nuts and bolts to protect them from detachment?
☐	☐	☐	12. Do all metal edges have rolled edging or rounded capping?
☐	☐	☐	13. Are metal materials painted, galvanized, anodized, or composed of non-rusting materials?
☐	☐	☐	14. When located in direct sunlight, have metal materials been coated in plastic to avoid the risk of a contact-burn injury? ▷ *Bare or painted metal surfaces should be avoided in intense, direct sunlight.*
☐	☐	☐	15. Are plastic materials ultraviolet-stabilized to resist fading? ▷ *This question is based on the authors' opinion and is not addressed by CPSC or ASTM.*

TO CONTINUE AUDIT, COMPLETE ANNUAL AND PERIODIC INSPECTIONS.

8 BRIDGES, CLATTER

PARK NAME　　　　　　　　　DATE OF INSPECTION　　　　INSPECTOR

Yes	No	N/A	**ANNUAL INSPECTION**
☐	☐	☐	16. Is the bridge free of head and neck entrapments (see Inspection Procedures)?
☐	☐	☐	17. Do protrusions meet the protrusion test criteria (see Inspection Procedures)?
☐	☐	☐	18. Is the bridge free of hollow support posts or tubes with open ends?
☐	☐	☐	19. Are equipment footings securely anchored?
☐	☐	☐	20. Are wood materials naturally rot- and insect-resistant, or treated with a wood preservative below and up to 6 inches (150 mm) above the surface of the play area?
			If a wood preservative was used, list the preservative's name: _____.
☐	☐	☐	21. Is the wood preservative safe for use in children's play areas, as specified by ASTM F 1487 standards?
☐	☐	☐	22. Are paints free of lead (0.06% maximum lead by dry weight) as specified by ASTM F 1487 standards?

TO CONTINUE ANNUAL INSPECTION, COMPLETE PERIODIC INSPECTION.

8 BRIDGES, CLATTER

PARK NAME _____ DATE OF INSPECTION _____ INSPECTOR _____

Yes	No	N/A	**PERIODIC INSPECTION**
☐	☐	☐	23. Is the bridge stable and without severe structural deterioration, such as at the footings and joints?
☐	☐	☐	24. Is the bridge free of loose, missing, or broken parts and vandalism?
☐	☐	☐	25. Is the bridge free of sharp points, corners, or edges?
☐	☐	☐	26. Are all moving suspended elements connected to the fixed support with bearings that reduce friction and wear? ▷ *A steel cable permanently affixed to a hanger assembly meets this requirement.*
☐	☐	☐	27. Are chains or cables without significant wear? ▷ *Wear is indicated by visible elongation, deformation, indentation, rust, or corrosion.*
☐	☐	☐	28. Are cables free of frayed or projecting wires?
☐	☐	☐	29. Are cables or chains fixed tightly at both ends so that there is no possibility of overlapping and entrapping a child?
☐	☐	☐	30. Is all hardware present, securely attached, and free of significant wear? ▷ *Wear is indicated by visible elongation, deformation, indentation, rust, corrosion, or stripping.*
☐	☐	☐	31. Do bolt ends extend no more than two threads beyond the face of the nut?
☐	☐	☐	32. Are all fastening devices closed to prevent entanglement (see Definitions)?
☐	☐	☐	33. Are wood materials free of warping, wood rot, insect damage, cupping, and checking?
☐	☐	☐	34. Are wood materials free of splinters, heart center, and loose or missing knots?
☐	☐	☐	35. Are metal materials free of rust, corrosion, peeling paint, and bent parts?
☐	☐	☐	36. Are plastic parts unbroken, unchipped, and uncracked, particularly at joints and connections?
☐	☐	☐	37. Is the bridge free of chipped, peeling, or worn paint?

©1997 MIG Communications

9 BRIDGES, STATIONARY

PARK NAME _____ DATE OF INSPECTION _____ INSPECTOR _____

Yes	No	N/A	AUDIT
☐	☐	☐	1. Does the bridge meet all standards for structural integrity as specified by ASTM F 1487?
☐	☐	☐	2. Does the bridge have a 72-inch (1800 mm) unobstructed use zone?
☐	☐	☐	3. Are vertical angles greater than 55 degrees? ▷ *Inverted angles or angles with a filled apex are exempt (see Definitions).*
☐	☐	☐	4. a. For 2- to 5-year-olds, are all play platforms that are more than 30 inches (760 mm) high enclosed by a protective barrier 29 inches (740 mm) or greater in height?
☐	☐	☐	b. For 5- to 12-year-olds, are all play platforms that are more than 48 inches (1200 mm) high enclosed by a protective barrier 38 inches (970 mm) or greater in height?
☐	☐	☐	5. a. For 2- to 5-year-olds, are all play equipment platforms over 20 inches (510 mm) high enclosed by a guardrail that is a maximum 23 inches (580 mm) high at the lower edge and 29 inches (740 mm) high at the top edge?
☐	☐	☐	b. For 5- to 12-year-olds, are all play equipment platforms over 30 inches (760 mm) high enclosed by a guardrail that is a maximum 28 inches (710 mm) high at the lower edge and 38 inches (970 mm) high at the top edge?

©1997 MIG Communications

9 BRIDGES, STATIONARY

| PARK NAME | DATE OF INSPECTION | INSPECTOR |

Yes	No	N/A		AUDIT (cont.)
☐	☐	☐	6.	Is the bridge free of extra holes that could harbor nesting insects? ▷ *This question is based on the authors' opinion and is not addressed by CPSC or ASTM.*
☐	☐	☐	7.	Is the bridge free of pinch, crush, and shear points (see Definitions)?
☐	☐	☐	8.	Is the bridge free of cables, wires, or other suspended hazards hung within 45 degrees of horizontal (see Definitions)?
☐	☐	☐	9.	Are lock washers, self-locking nuts, or other locking means provided for all nuts and bolts to protect them from detachment?
☐	☐	☐	10.	Do all metal edges have rolled edging or rounded capping?
☐	☐	☐	11.	Are metal materials painted, galvanized, anodized, or composed of non-rusting materials?
☐	☐	☐	12.	When located in direct sunlight, have metal materials been coated in plastic to avoid the risk of a contact-burn injury? ▷ *Bare or painted metal surfaces should be avoided in intense, direct sunlight.*
☐	☐	☐	13.	Are plastic materials ultraviolet-stabilized to resist fading? ▷ *This question is based on the authors' opinion and is not addressed by CPSC or ASTM.*

TO CONTINUE AUDIT, COMPLETE ANNUAL AND PERIODIC INSPECTIONS.

9 BRIDGES, STATIONARY

PARK NAME | DATE OF INSPECTION | INSPECTOR

Yes	No	N/A	**ANNUAL INSPECTION**
☐	☐	☐	14. Is the bridge free of head and neck entrapments (see Inspection Procedures)?
☐	☐	☐	15. Do protrusions meet the protrusion test criteria (See Inspection Procedures)?
☐	☐	☐	16. Is the bridge free of hollow support posts or tubes with open ends?
☐	☐	☐	17. Are equipment footings securely anchored?
☐	☐	☐	18. Are wood materials naturally rot- and insect-resistant, or treated with a wood preservative below and up to 6 inches (150 mm) above the surface of the play area?

If a wood preservative was used, list the preservative's name:

_____.

Yes	No	N/A	
☐	☐	☐	19. Is the wood preservative safe for use in children's play areas, as specified by ASTM F 1487 standards?
☐	☐	☐	20. Are paints free of lead (0.06% maximum lead by dry weight) as specified by ASTM F 1487 standards?

TO CONTINUE ANNUAL INSPECTION, COMPLETE PERIODIC INSPECTION.

©1997 MIG Communications

9 BRIDGES, STATIONARY

PARK NAME   DATE OF INSPECTION   INSPECTOR

Yes	No	N/A	PERIODIC INSPECTION
☐	☐	☐	21. Is the bridge stable and without severe structural deterioration, such as at the footings and joints?
☐	☐	☐	22. Is the bridge free of loose, missing, or broken parts and vandalism?
☐	☐	☐	23. Is the bridge free of sharp points, corners, or edges?
☐	☐	☐	24. Is all hardware present, securely attached, and free of significant wear? ▷ *Wear is indicated by visible elongation, deformation, indentation, rust, corrosion, or stripping.*
☐	☐	☐	25. Do bolt ends extend no more than two threads beyond the face of the nut?
☐	☐	☐	26. Are all fastening devices closed to prevent entanglement (see Definitions)?
☐	☐	☐	27. Are wood materials free of warping, wood rot, insect damage, cupping, and checking?
☐	☐	☐	28. Are wood materials free of splinters, heart center, and loose or missing knots?
☐	☐	☐	29. Are metal materials free of rust, corrosion, peeling paint, and bent parts?
☐	☐	☐	30. Are plastic parts unbroken, unchipped, and uncracked, particularly at joints and connections?
☐	☐	☐	31. Is the bridge free of chipped, peeling, or worn paint?

©1997 MIG Communications

10 CLIMBERS

PARK NAME _____ DATE OF INSPECTION _____ INSPECTOR _____

Note: According to CPSC, arch climbers are not recommended for children under 4.

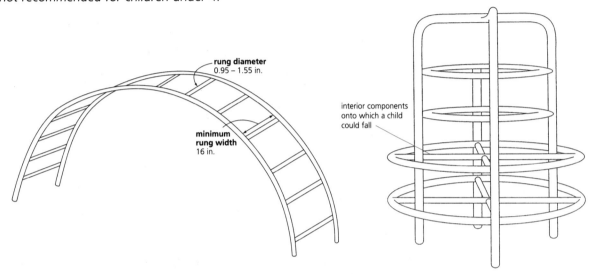

Yes	No	N/A	AUDIT
☐	☐	☐	1. Does the climber meet all standards for structural integrity as specified by ASTM F 1487?
☐	☐	☐	2. Does the climber have a 72-inch (1800 mm) unobstructed use zone?
☐	☐	☐	3. Are vertical angles greater than 55 degrees? ▷ *Inverted angles or angles with a filled apex are exempt (see Definitions).*
☐	☐	☐	4. Is the climber free of extra holes that could harbor nesting insects? ▷ *This question is based on the authors' opinion and is not addressed by CPSC or ASTM.*
☐	☐	☐	5. Is the climber free of pinch, crush, and shear points (see Definitions)?
☐	☐	☐	6. Is the climber free of cables, wires, or other suspended hazards hung within 45 degrees of horizontal (see Definitions)?
☐	☐	☐	7. Do the rungs measure between 0.95 and 1.55 inches (24.1 and 39.4 mm) in diameter?
☐	☐	☐	8. Are the rungs at least 16 inches (400 mm) wide?

©1997 MIG Communications

10 CLIMBERS

PARK NAME DATE OF INSPECTION INSPECTOR

Yes	No	N/A	**AUDIT (cont.)**
☐	☐	☐	9. Are the rungs securely attached so that they do not rotate?
☐	☐	☐	10. Are the rungs evenly spaced?
☐	☐	☐	11. Is the climber free of components in its interior onto which a child may fall from a height greater than 18 inches (450 mm), as recommended by CPSC?
☐	☐	☐	12. When attached to a composite structure, is there a less challenging means of entry and exit in addition to climber access to the structure?
☐	☐	☐	13. Does the climber provide hand support during climbing and at the transition where the climber connects to the composite structure platform?
☐	☐	☐	14. Are lock washers, self-locking nuts, or other locking means provided for all nuts and bolts to protect them from detachment?
☐	☐	☐	15. Do all metal edges have rolled edging or rounded capping?
☐	☐	☐	16. Are metal materials painted, galvanized, anodized, or composed of non-rusting materials?
☐	☐	☐	17. When located in direct sunlight, have metal materials been coated in plastic to avoid the risk of a contact-burn injury? ▷ *Bare or painted metal surfaces should be avoided in intense, direct sunlight.*
☐	☐	☐	18. Are plastic materials ultraviolet-stabilized to resist fading? ▷ *This question is based on the authors' opinion and is not addressed by CPSC or ASTM.*

TO CONTINUE AUDIT, COMPLETE ANNUAL AND PERIODIC INSPECTIONS.

10 CLIMBERS

PARK NAME   **DATE OF INSPECTION**   **INSPECTOR**

Yes	No	N/A	**ANNUAL INSPECTION**
☐	☐	☐	19. Is the climber free of head and neck entrapments (see Inspection Procedures)?
☐	☐	☐	20. Do protrusions meet the protrusion test criteria (see Inspection Procedures)?
☐	☐	☐	21. Is the climber free of hollow support posts or tubes with open ends?
☐	☐	☐	22. Are equipment footings securely anchored?
☐	☐	☐	23. Are wood materials naturally rot- and insect-resistant, or treated with a wood preservative below and up to 6 inches (150 mm) above the surface of the play area? If a wood preservative was used, list the preservative's name: _____.
☐	☐	☐	24. Is the wood preservative safe for use in children's play areas, as specified by ASTM F 1487 standards?
☐	☐	☐	25. Are paints free from lead (0.06% maximum lead by dry weight) as specified by ASTM F 1487 standards?

TO CONTINUE ANNUAL INSPECTION, COMPLETE PERIODIC INSPECTION.

10 CLIMBERS

PARK NAME _____ DATE OF INSPECTION _____ INSPECTOR _____

Yes	No	N/A	**PERIODIC INSPECTION**
☐	☐	☐	26. Is the climber stable and without severe structural deterioration, such as at the footings and joints?
☐	☐	☐	27. Is the climber free of loose, missing, or broken parts and vandalism?
☐	☐	☐	28. Is the climber free of sharp points, corners, or edges?
☐	☐	☐	29. Is all hardware present, securely attached, and free of significant wear? ▷ *Wear is indicated by visible elongation, deformation, indentation, rust, corrosion, or stripping.*
☐	☐	☐	30. Do bolt ends extend no more than two threads beyond the face of the nut?
☐	☐	☐	31. Are all fastening devices closed to prevent entanglement (see Definitions)?
☐	☐	☐	32. Are wood materials free of warping, wood rot, insect damage, cupping, and checking?
☐	☐	☐	33. Are wood materials free of splinters, heart center, and loose or missing knots?
☐	☐	☐	34. Are metal materials free of rust, corrosion, peeling paint, and bent parts?
☐	☐	☐	35. Are plastic parts unbroken, unchipped, and uncracked, particularly at joints and connections?
☐	☐	☐	36. Is the climber free of chipped, peeling, or worn paint?

©1997 MIG Communications

11 CLIMBERS, FLEXIBLE

PARK NAME _____   DATE OF INSPECTION _____   INSPECTOR _____

Note: In the authors' opinion, flexible climbers are not recommended for children under 3.

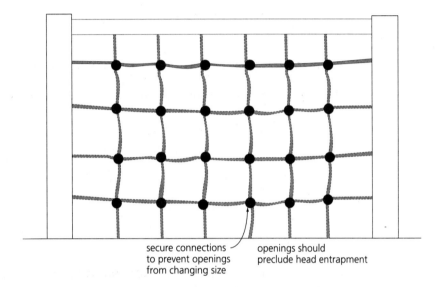

secure connections to prevent openings from changing size

openings should preclude head entrapment

Yes	No	N/A	**AUDIT**
☐	☐	☐	1. Does the climber meet all standards for structural integrity as specified by ASTM F 1487?
☐	☐	☐	2. Does the climber have a 72-inch (1800 mm) unobstructed use zone?
☐	☐	☐	3. Are vertical angles greater than 55 degrees? ▷ *Inverted angles or angles with a filled apex are exempt (see Definitions).*
☐	☐	☐	4. For 3- to 5-year-olds, does the climber allow users to bring both feet to the same level before ascending to the next level?
☐	☐	☐	5. When the climber is used to provide access to a composite structure, is another means of access also provided?
☐	☐	☐	6. Is the climber free of extra holes that could harbor nesting insects? ▷ *This question is based on the authors' opinion and is not addressed by CPSC or ASTM.*
☐	☐	☐	7. Is the climber free of pinch, crush, and shear points (see Definitions)?
☐	☐	☐	8. Is the climber free of cables, wires, or other suspended hazards hung within 45 degrees of horizontal (see Definitions)?

©1997 MIG Communications

11 CLIMBERS, FLEXIBLE

PARK NAME   DATE OF INSPECTION   INSPECTOR

Yes	No	N/A	AUDIT (cont.)
☐	☐	☐	9. Do chains or cables meet ASTM F 1487 structural integrity requirements?
☐	☐	☐	10. Do cables measure at least 1 inch (25 mm) in diameter?
☐	☐	☐	11. Are lock washers, self-locking nuts, or other locking means provided for all nuts and bolts to protect them from detachment?
☐	☐	☐	12. Do all metal edges have rolled edging or rounded capping?
☐	☐	☐	13. Are metal materials painted, galvanized, anodized, or composed of non-rusting materials?
☐	☐	☐	14. When located in direct sunlight, have metal materials been coated in plastic to avoid the risk of a contact-burn injury? ▷ *Bare or painted metal surfaces should be avoided in intense, direct sunlight.*
☐	☐	☐	15. Are plastic materials ultraviolet-stabilized to resist fading? ▷ *This question is based on the authors' opinion and is not addressed by CPSC or ASTM.*

TO CONTINUE AUDIT, COMPLETE ANNUAL AND PERIODIC INSPECTIONS.

11 CLIMBERS, FLEXIBLE

PARK NAME _____ DATE OF INSPECTION _____ INSPECTOR _____

Yes	No	N/A	**ANNUAL INSPECTION**
☐	☐	☐	16. Is the climber free of head and neck entrapments (see Inspection Procedures)?
☐	☐	☐	17. Do protrusions meet the protrusion test criteria (see Inspection Procedures)?
☐	☐	☐	18. Is the climber free of hollow support posts or tubes with open ends?
☐	☐	☐	19. Are equipment footings securely anchored?
☐	☐	☐	20. Are wood materials naturally rot- and insect-resistant, or treated with a wood preservative below and up to 6 inches (150 mm) above the surface of the play area?

If a wood preservative was used, list the preservative's name:

_____.

Yes	No	N/A	
☐	☐	☐	21. Is the wood preservative safe for use in children's play areas, as specified by ASTM F 1487 standards?
☐	☐	☐	22. Are paints free from lead (0.06% maximum lead by dry weight) as specified by ASTM F 1487 standards?

TO CONTINUE ANNUAL INSPECTION, COMPLETE PERIODIC INSPECTION.

11 CLIMBERS, FLEXIBLE

| PARK NAME | DATE OF INSPECTION | INSPECTOR |

Yes	No	N/A	**PERIODIC INSPECTION**
☐	☐	☐	23. Is the climber stable and without severe structural deterioration, such as at the footings and joints?
☐	☐	☐	24. Is the climber free of loose, missing, or broken parts and vandalism?
☐	☐	☐	25. Is the climber free of sharp points, corners, or edges?
☐	☐	☐	26. Is the climber adjusted to eliminate loose cable?
☐	☐	☐	27. Are connections securely fixed to prevent net openings from changing size?
☐	☐	☐	28. Are chains or cables without significant wear? ▷ *Wear is indicated by visible elongation, deformation, indentation, rust, or corrosion.*
☐	☐	☐	29. Are cables free of frayed or projecting wires?
☐	☐	☐	30. Are cables or chains fixed tightly at both ends so that there is no possibility of overlapping and entrapping a child?
☐	☐	☐	31. When one end of the flexible climber is attached at ground level, is the anchoring device below the playing surface?
☐	☐	☐	32. Is all hardware present, securely attached, and free of significant wear? ▷ *Wear is indicated by visible elongation, deformation, indentation, rust, corrosion, or stripping.*
☐	☐	☐	33. Do bolt ends extend no more than two threads beyond the face of the nut?
☐	☐	☐	34. Are all fastening devices closed to prevent entanglement (see Definitions)?
☐	☐	☐	35. Are wood materials free of warping, wood rot, insect damage, cupping, and checking?
☐	☐	☐	36. Are wood materials free of splinters, heart center, and loose or missing knots?
☐	☐	☐	37. Are metal materials free of rust, corrosion, peeling paint, and bent parts?
☐	☐	☐	38. Are plastic parts unbroken, unchipped, and uncracked, particularly at joints and connections?
☐	☐	☐	39. Is the climber free of chipped, peeling, or worn paint?

©1997 MIG Communications

12 FIRE POLES

1 OF 4

PARK NAME DATE OF INSPECTION INSPECTOR

Note: In the authors' opinion, fire poles are not recommended for children under 5.

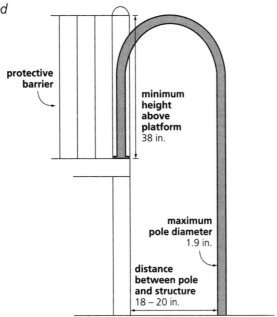

Yes	No	N/A	**AUDIT**
☐	☐	☐	1. Does the fire pole meet all standards for structural integrity as specified by ASTM F 1487?
☐	☐	☐	2. Does the fire pole have a 72-inch (1800 mm) unobstructed use zone?
☐	☐	☐	3. Are vertical angles greater than 55 degrees? ▷ *Inverted angles or angles with a filled apex are exempt (see Definitions).*
☐	☐	☐	4. Is the fire pole attached to a composite structure platform with a maximum height of 72 inches (1800 mm)? ▷ *A fire pole should not be installed as freestanding equipment.*
☐	☐	☐	5. Is the fire pole free of extra holes that could harbor nesting insects? ▷ *This question is based on the authors' opinion and is not addressed by CPSC or ASTM.*
☐	☐	☐	6. Is the fire pole free of pinch, crush, and shear points (see Definitions)?
☐	☐	☐	7. Is the fire pole free of cables, wires, or other suspended hazards hung within 45 degrees of horizontal (see Definitions)?

12 FIRE POLES

PARK NAME   DATE OF INSPECTION   INSPECTOR

Yes	No	N/A	AUDIT (cont.)
☐	☐	☐	8. Does the fire pole measure no more than 1.9 inches (48 mm) in diameter?
☐	☐	☐	9. Is the sliding surface continuous and smooth with no protruding bolts or seams?
☐	☐	☐	10. Is the fire pole free of changes in direction along the sliding portion?
☐	☐	☐	11. Do rails support a child's transition to sliding?
☐	☐	☐	12. Is access to the sliding pole provided at only one location?
☐	☐	☐	13. Does the fire pole extend at least 38 inches (960 mm) above the level of the access platform?
☐	☐	☐	14. Is the distance between the fire pole and the composite structure between 18 and 20 inches (460 and 510 mm) along the entire length of the pole?
☐	☐	☐	15. Are lock washers, self-locking nuts, or other locking means provided for all nuts and bolts to protect them from detachment?
☐	☐	☐	16. Do all metal edges have rolled edging or rounded capping?
☐	☐	☐	17. Are metal materials painted, galvanized, anodized, or composed of non-rusting materials?
☐	☐	☐	18. When located in direct sunlight, have metal materials been coated in plastic to avoid the risk of a contact-burn injury? ▷ *Bare or painted metal surfaces should be avoided in intense, direct sunlight.*
☐	☐	☐	19. Are plastic materials ultraviolet-stabilized to resist fading? ▷ *This question is based on the authors' opinion and is not addressed by CPSC or ASTM.*

TO CONTINUE AUDIT, COMPLETE ANNUAL AND PERIODIC INSPECTIONS.

12 FIRE POLES

PARK NAME _____ DATE OF INSPECTION _____ INSPECTOR _____

Yes	No	N/A	**ANNUAL INSPECTION**
☐	☐	☐	20. Is the fire pole free of head and neck entrapments (see Inspection Procedures)?
☐	☐	☐	21. Do protrusions meet the protrusion test criteria (see Inspection Procedures)?
☐	☐	☐	22. Is the fire pole free of hollow support posts or tubing with open ends?
☐	☐	☐	23. Are equipment footings securely anchored?
☐	☐	☐	24. Are wood materials naturally rot- and insect-resistant, or treated with a wood preservative below and up to 6 inches (150 mm) above the surface of the play area? If a wood preservative was used, list the preservative's name: _____ .
☐	☐	☐	25. Is the wood preservative safe for use in children's play areas, as specified by ASTM F 1487 standards?
☐	☐	☐	26. Are paints free from lead (0.06% maximum lead by dry weight) as specified by ASTM F 1487 standards?

TO CONTINUE ANNUAL INSPECTION, COMPLETE PERIODIC INSPECTION.

12 FIRE POLES

| PARK NAME | DATE OF INSPECTION | INSPECTOR |

Yes	No	N/A	**PERIODIC INSPECTION**
☐	☐	☐	27. Is the fire pole stable and without severe structural deterioration, such as at the footings and joints?
☐	☐	☐	28. Is the fire pole free of loose, missing, or broken parts and vandalism?
☐	☐	☐	29. Is the fire pole free of sharp points, corners, or edges?
☐	☐	☐	30. Is all hardware present, securely attached, and free of significant wear? ▷ *Wear is indicated by visible elongation, deformation, indentation, rust, corrosion, or stripping.*
☐	☐	☐	31. Do bolt ends extend no more than two threads beyond the face of the nut?
☐	☐	☐	32. Are all fastening devices closed to prevent entanglement (see Definitions)?
☐	☐	☐	33. Are wood materials free of warping, wood rot, insect damage, cupping, and checking?
☐	☐	☐	34. Are wood materials free of splinters, heart center, and loose or missing knots?
☐	☐	☐	35. Are metal materials free of rust, corrosion, peeling paint, and bent parts?
☐	☐	☐	36. Are plastic parts unbroken, unchipped, and uncracked, particularly at joints and connections?
☐	☐	☐	37. Is the fire pole free of chipped, peeling, or worn paint?

©1997 MIG Communications

13 HORIZONTAL LADDERS & RING TREKS 1 OF 4

PARK NAME　　　　　DATE OF INSPECTION　　　　　INSPECTOR

Note: According to CPSC, horizontal ladders and ring treks are not recommended for children under 4 years.

horizontal ladder
- maximum distance 10 in.
- maximum distance between rungs 15 in.
- maximum height 5 to 12 years: 84 in.
- maximum platform height 36 in.

ring trek
- maximum height 84 in.
- maximum platform height 36 in.

Yes	No	N/A	**AUDIT**
☐	☐	☐	1. Does the equipment meet all standards for structural integrity as specified by ASTM F 1487?
☐	☐	☐	2. Does the equipment have a 72-inch (1800 mm) unobstructed use zone?
☐	☐	☐	3. Are vertical angles greater than 55 degrees? ▷ *Inverted angles or angles with a filled apex are exempt (see Definitions).*
☐	☐	☐	4. For 5- to 12-year-olds, is the equipment no more than 84 inches (2100 mm) high?
☐	☐	☐	5. Are the takeoff and landing platforms no more than 36 inches (910 mm) high?
☐	☐	☐	6. Is the equipment free of extra holes that could harbor nesting insects? ▷ *This question is based on the authors' opinion and is not addressed by CPSC or ASTM.*
☐	☐	☐	7. Is the equipment free of pinch, crush, and shear points (see Definitions)?

©1997 MIG Communications

13 HORIZONTAL LADDERS & RING TREKS

PARK NAME _____ DATE OF INSPECTION _____ INSPECTOR _____

Yes	No	N/A	AUDIT (cont.)
☐	☐	☐	8. Is the equipment free of cables, wires, or other suspended hazards hung within 45 degrees of horizontal (see Definitions)?
☐	☐	☐	9. Do the rungs measure between 0.95 and 1.55 inches (24.1 and 39.4 mm) in diameter?
☐	☐	☐	10. Are the rungs spaced no more than 15 inches (380 mm) apart?
☐	☐	☐	11. Are the first handholds on either ends of the equipment inset so that they are not located directly above the platform or rungs used for equipment or exit?
☐	☐	☐	12. Is the horizontal distance between the first handhold and the takeoff or landing platforms no more than 10 inches (250 mm)?
☐	☐	☐	13. When access and egress are provided by means of rungs, is the horizontal distance from the takeoff and landing platforms to the first handhold between 8 and 10 inches (200 and 250 mm)?
☐	☐	☐	14. Is the maximum distance from the top surface of the access and egress points to the first handhold 45 inches (1140 mm)?
☐	☐	☐	15. Do chains or cables meet ASTM F 1487 structural integrity requirements?
☐	☐	☐	16. Do cables measure at least 1 inch (25 mm) in diameter?
☐	☐	☐	17. Are lock washers, self-locking nuts, or other locking means provided for all nuts and bolts to protect them from detachment?
☐	☐	☐	18. Do all metal edges have rolled edging or rounded capping?
☐	☐	☐	19. Are metal materials painted, galvanized, anodized, or composed of non-rusting materials?
☐	☐	☐	20. When located in direct sunlight, have metal materials been coated in plastic to avoid the risk of a contact-burn injury? ▷ *Bare or painted metal surfaces should be avoided in intense, direct sunlight.*
☐	☐	☐	21. Are plastic materials ultraviolet-stabilized to resist fading? ▷ *This question is based on the authors' opinion and is not addressed by CPSC or ASTM.*

TO CONTINUE AUDIT, COMPLETE ANNUAL AND PERIODIC INSPECTIONS.

13 HORIZONTAL LADDERS & RING TREKS

PARK NAME   DATE OF INSPECTION   INSPECTOR

Yes	No	N/A	
			ANNUAL INSPECTION
☐	☐	☐	22. Is the equipment free of head and neck entrapments (see Inspection Procedures)?
☐	☐	☐	23. Do protrusions meet the protrusion test criteria (see Inspection Procedures)?
☐	☐	☐	24. Is the equipment free of hollow support posts or tubes with open ends?
☐	☐	☐	25. Are equipment footings securely anchored?
☐	☐	☐	26. Are wood materials naturally rot- and insect-resistant, or treated with a wood preservative below and up to 6 inches (150 mm) above the surface of the play area?
			If a wood preservative was used, list the preservative's name: _____.
☐	☐	☐	27. Is the wood preservative safe for use in children's play areas, as specified by ASTM F 1487 standards?
☐	☐	☐	28. Are paints free of lead (0.06% maximum lead by dry weight) as specified by ASTM F 1487 standards?

TO CONTINUE ANNUAL INSPECTION, COMPLETE PERIODIC INSPECTION.

13 HORIZONTAL LADDERS & RING TREKS

PARK NAME   DATE OF INSPECTION   INSPECTOR

Yes	No	N/A	**PERIODIC INSPECTION**
☐	☐	☐	29. Is the equipment stable and without severe structural deterioration, such as at the footings and joints?
☐	☐	☐	30. Is the equipment free of loose, missing, or broken parts and vandalism?
☐	☐	☐	31. Is the equipment free of sharp points, corners, or edges?
☐	☐	☐	32. Are the rungs securely fixed so that they do not rotate?
☐	☐	☐	33. Are chains or cables without significant wear? ▷ *Wear is indicated by visible elongation, deformation, indentation, rust, or corrosion.*
☐	☐	☐	34. Are cables free of frayed or projecting wires?
☐	☐	☐	35. Is all hardware present, securely attached, and free of significant wear? ▷ *Wear is indicated by visible elongation, deformation, indentation, rust, corrosion, or stripping.*
☐	☐	☐	36. Do bolt ends extend no more than two threads beyond the face of the nut?
☐	☐	☐	37. Are all fastening devices closed to prevent entanglement (see Definitions)?
☐	☐	☐	38. Are wood materials free of warping, wood rot, insect damage, cupping, and checking?
☐	☐	☐	39. Are wood materials free of splinters, heart center, and loose or missing knots?
☐	☐	☐	40. Are metal materials free of rust, corrosion, peeling paint, and bent parts?
☐	☐	☐	41. Are plastic parts unbroken, unchipped, and uncracked, particularly at joints and connections?
☐	☐	☐	42. Is the equipment free of chipped, peeling, or worn paint?

©1997 MIG Communications

14 PLAYHOUSES

PARK NAME   DATE OF INSPECTION   INSPECTO

Yes	No	N/A	AUDIT
☐	☐	☐	1. Does the playhouse meet all standards for structural integrity as specified by ASTM F 1487?
☐	☐	☐	2. Does the playhouse have a 72-inch (1800 mm) unobstructed use zone? ▷ *Two nonclimbable playhouses may have overlapping use zones (see Definitions).*
☐	☐	☐	3. Are vertical angles greater than 55 degrees? ▷ *Inverted angles or angles with a filled apex are exempt (see Definitions).*
☐	☐	☐	4. Is the playhouse free of extra holes that could harbor nesting insects? ▷ *This question is based on the authors' opinion and is not addressed by CPSC or ASTM.*
☐	☐	☐	5. Is the playhouse free of pinch, crush, and shear points (see Definitions)?
☐	☐	☐	6. Is the playhouse free of cables, wires, or other suspended hazards hung within 45 degrees of horizontal (see Definitions)?

©1997 MIG Communications

14 PLAYHOUSES

PARK NAME   DATE OF INSPECTION   INSPECTOR

Yes	No	N/A	**AUDIT (cont.)**
☐	☐	☐	7. Are there clear sight-lines into the playhouse from more than one location? ▷ *This question is based on the authors' opinion and is not addressed by CPSC or ASTM.*
☐	☐	☐	8. Are lock washers, self-locking nuts, or other locking means provided for all nuts and bolts to protect them from detachment?
☐	☐	☐	9. Do all metal edges have rolled edging or rounded capping?
☐	☐	☐	10. Are metal materials painted, galvanized, anodized, or composed of non-rusting materials?
☐	☐	☐	11. When located in direct sunlight, have metal materials been coated in plastic to avoid the risk of a contact-burn injury? ▷ *Bare or painted metal surfaces should be avoided in intense, direct sunlight.*
☐	☐	☐	12. Are plastic materials ultraviolet-stabilized to resist fading? ▷ *This question is based on the authors' opinion and is not addressed by CPSC or ASTM.*

TO CONTINUE AUDIT, COMPLETE ANNUAL AND PERIODIC INSPECTIONS.

14 PLAYHOUSES

PARK NAME   DATE OF INSPECTION   INSPECTOR

Yes	No	N/A	**ANNUAL INSPECTION**
☐	☐	☐	13. Is the playhouse free of head and neck entrapments (see Inspection Procedures)?
☐	☐	☐	14. Do protrusions meet the protrusion test criteria (see Inspection Procedures)?
☐	☐	☐	15. Is the playhouse free of hollow support posts or tubes with open ends?
☐	☐	☐	16. Are equipment footings securely anchored?
☐	☐	☐	17. Are wood materials naturally rot- and insect-resistant, or treated with a wood preservative below and up to 6 inches (150 mm) above the surface of the play area? If a wood preservative was used, list the preservative's name: _____ .
☐	☐	☐	18. Is the wood preservative safe for use in children's play areas, as specified by ASTM F 1487 standards?
☐	☐	☐	19. Are paints free of lead (0.06% maximum lead by dry weight) as specified by ASTM F 1487 standards?

TO CONTINUE ANNUAL INSPECTION, COMPLETE PERIODIC INSPECTION.

14 PLAYHOUSES

PARK NAME   **DATE OF INSPECTION**   **INSPECTOR**

Yes	No	N/A	**PERIODIC INSPECTION**
☐	☐	☐	20. Is the playhouse stable and without severe structural deterioration, such as at the footings and joints?
☐	☐	☐	21. Is the playhouse free of loose, missing, or broken parts and vandalism?
☐	☐	☐	22. Is the playhouse free of sharp points, corners, or edges?
☐	☐	☐	23. Is all hardware present, securely attached, and free of significant wear? ▷ *Wear is indicated by visible elongation, deformation, indentation, rust, corrosion, or stripping.*
☐	☐	☐	24. Do bolt ends extend no more than two threads beyond the face of the nut?
☐	☐	☐	25. Are all fastening devices closed to prevent entanglement (see Definitions)?
☐	☐	☐	26. Are wood materials free of warping, wood rot, insect damage, cupping, and checking?
☐	☐	☐	27. Are wood materials free of splinters, heart center, and loose or missing knots?
☐	☐	☐	28. Are metal materials free of rust, corrosion, peeling paint, and bent parts?
☐	☐	☐	29. Are plastic parts unbroken, unchipped, and uncracked, particularly at joints and connections?
☐	☐	☐	30. Is the playhouse free of chipped, peeling, or worn paint?

©1997 MIG Communications

15 SLIDES

PARK NAME   **DATE OF INSPECTION**   **INSPECTOR**

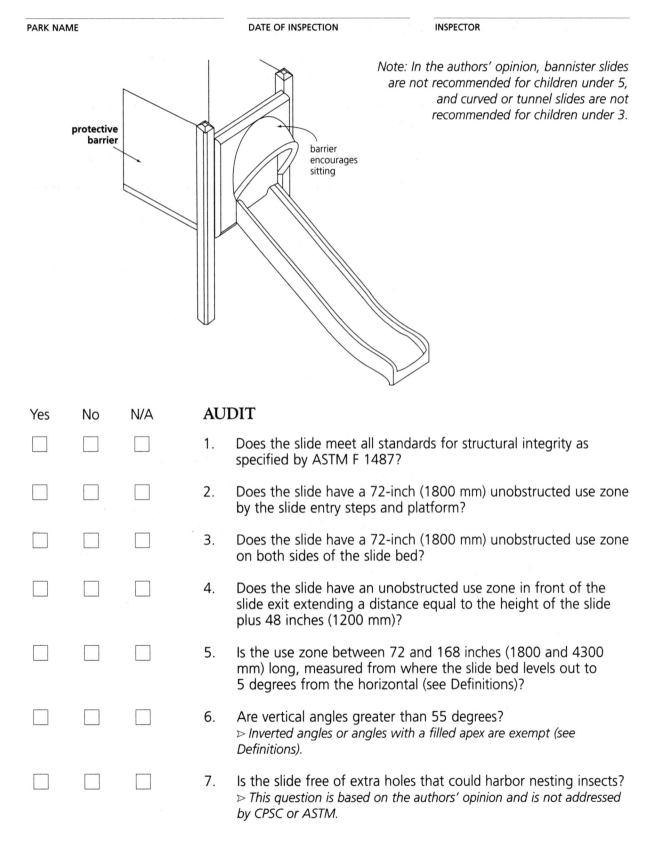

Note: In the authors' opinion, bannister slides are not recommended for children under 5, and curved or tunnel slides are not recommended for children under 3.

Yes	No	N/A	**AUDIT**
☐	☐	☐	1. Does the slide meet all standards for structural integrity as specified by ASTM F 1487?
☐	☐	☐	2. Does the slide have a 72-inch (1800 mm) unobstructed use zone by the slide entry steps and platform?
☐	☐	☐	3. Does the slide have a 72-inch (1800 mm) unobstructed use zone on both sides of the slide bed?
☐	☐	☐	4. Does the slide have an unobstructed use zone in front of the slide exit extending a distance equal to the height of the slide plus 48 inches (1200 mm)?
☐	☐	☐	5. Is the use zone between 72 and 168 inches (1800 and 4300 mm) long, measured from where the slide bed levels out to 5 degrees from the horizontal (see Definitions)?
☐	☐	☐	6. Are vertical angles greater than 55 degrees? ▷ *Inverted angles or angles with a filled apex are exempt (see Definitions).*
☐	☐	☐	7. Is the slide free of extra holes that could harbor nesting insects? ▷ *This question is based on the authors' opinion and is not addressed by CPSC or ASTM.*

©1997 MIG Communications

15 SLIDES

| PARK NAME | DATE OF INSPECTION | INSPECTOR |

Yes	No	N/A	**AUDIT (cont.)**
☐	☐	☐	8. Is the slide free of pinch, crush, and shear points (see Definitions)?
☐	☐	☐	9. Is the slide free of cables, wires, or other suspended hazards hung within 45 degrees of horizontal (see Definitions)?
☐	☐	☐	10. Is the slope of the slide 50 degrees or less?
☐	☐	☐	11. Does the slide surface have a height-length ratio that does not exceed 0.577 (see Definitions)?
☐	☐	☐	12. Is the platform at the slide entrance at least 22 inches (560 mm) in length and as wide as the slide chute?
☐	☐	☐	13. a. For 2- to 5-year-olds, are slide platforms more than 30 inches (760 mm) in height surrounded by protective barriers at least 29 inches (740 mm) high?
☐	☐	☐	b. For 5- to 12-year-olds, are slide platforms more than 48 inches (1200 mm) in height surrounded by protective barriers at least 38 inches (970 mm) high?
☐	☐	☐	14. a. For 2- to 5-year-olds, are all play equipment platforms over 20 inches (510 mm) high enclosed by a guardrail that is a maximum 23 inches (580 mm) high at the lower edge and 29 inches (740 mm) high at the top edge?
☐	☐	☐	b. For 5- to 12-year-olds, are all play equipment platforms over 30 inches (760 mm) high enclosed by a guardrail that is a maximum 28 inches (710 mm) high at the lower edge and 38 inches (970 mm) high at the top edge?
☐	☐	☐	15. Are handrails provided at the slide entrance to facilitate the transition from standing to sitting?
☐	☐	☐	16. Is there a barrier at the slide entrance, such as a hood or a rail, to encourage the user to assume a sitting position?
☐	☐	☐	17. a. For 2- to 5-year-olds, is the inside width of the slide chute at least 12 inches (300 mm)?
☐	☐	☐	b. For 5- to 12-year-olds, is the inside width of the slide chute at least 16 inches (410 mm)?
☐	☐	☐	18. Are slide chute sidewalls at least 4 inches (100 mm) high?
☐	☐	☐	19. Do slide chute sidewalls extend along both sides of the chute for the entire length of the inclined sliding surface?
☐	☐	☐	20. Are no gaps present between the sidewalls and the surface?

©1997 MIG Communications

15 SLIDES

| PARK NAME | DATE OF INSPECTION | INSPECTOR |

Yes	No	N/A	**AUDIT (cont.)**
☐	☐	☐	21. For tube or tunnel slides, is the interior diameter at least 23 inches (580 mm)?
☐	☐	☐	22. For bannister slides, is the maximum pole diameter 1.9 inches (48 mm)?
☐	☐	☐	23. Is the slide exit region essentially horizontal and parallel to the ground? ▷ *The exit region should have a slope between 0 and -4 degrees.*
☐	☐	☐	24. Is the slide exit region at least 11 inches (280 mm) long?
☐	☐	☐	25. a. For slides no more than 48 inches (1200 mm) in height, is the exit region no more than 11 inches (280 mm) above the safety surface?
☐	☐	☐	b. For slides more than 48 inches (1200 mm) in height, is the exit region between 7 and 15 inches (180 and 380 mm) above the safety surface?
☐	☐	☐	26. Is the radius of curvature of the sliding surface in the exit region at least 30 inches (760 mm)?
☐	☐	☐	27. Are slide exit edges rounded or curved?
☐	☐	☐	28. a. Does the slide chute have a clearance zone that extends through the slide exit region, 60 inches (1500 mm) above the chute surface, and 21 inches (530 mm) to each side of the chute as measured from the inside face of the sidewalls (see Definitions)?
☐	☐	☐	b. For spiral slides, is the clearance zone 21 inches (530 mm) wide along the outer edge of the slide for the entire length of the slide, including the slide exit?
☐	☐	☐	29. Are lock washers, self-locking nuts, or other locking means provided for all nuts and bolts to protect them from detachment?
☐	☐	☐	30. Do all metal edges have rolled edging or rounded capping?
☐	☐	☐	31. Are metal materials painted, galvanized, anodized, or composed of non-rusting materials?
☐	☐	☐	32. When located in direct sunlight, have metal materials been coated in plastic to avoid the risk of a contact-burn injury? ▷ *Bare or painted metal surfaces should be avoided in intense, direct sunlight. Slide beds cannot be plastic-coated metal.*
☐	☐	☐	33. Are plastic materials ultraviolet-stabilized to resist fading? ▷ *This question is based on the authors' opinion and is not addressed by CPSC or ASTM.*

TO CONTINUE AUDIT, COMPLETE ANNUAL AND PERIODIC INSPECTIONS.

15 SLIDES

PARK NAME　　　　　　　　DATE OF INSPECTION　　　　INSPECTOR

Yes	No	N/A	**ANNUAL INSPECTION**
☐	☐	☐	34. Is the slide free of head and neck entrapments (see Inspection Procedures)?
☐	☐	☐	35. Do protrusions meet the protrusion test criteria (see Inspection Procedures)?
☐	☐	☐	36. Is the slide free of hollow support posts or tubes with open ends?
☐	☐	☐	37. Are equipment footings securely anchored?
☐	☐	☐	38. Are wood materials naturally rot- and insect-resistant, or treated with a wood preservative below and up to 6 inches (150 mm) above the surface of the play area?

If a wood preservative was used, list the preservative's name:

_____.

Yes	No	N/A	
☐	☐	☐	39. Is the wood preservative safe for use in children's play areas, as specified by ASTM F 1487 standards?
☐	☐	☐	40. Are paints free of lead (0.06% maximum lead by dry weight) as specified by ASTM F 1487 standards?

TO CONTINUE ANNUAL INSPECTION, COMPLETE PERIODIC INSPECTION.

15 SLIDES

PARK NAME _____ **DATE OF INSPECTION** _____ **INSPECTOR** _____

Yes	No	N/A	**PERIODIC INSPECTION**
☐	☐	☐	41. Is the slide stable and without severe structural deterioration, such as at the footings and joints?
☐	☐	☐	42. Is the slide free of loose, missing, or broken parts and vandalism?
☐	☐	☐	43. Is the slide free of sharp points, corners, or edges?
☐	☐	☐	44. Is the slide bed securely attached?
☐	☐	☐	45. For metal slide beds, is the slide located in a shaded area or oriented in a northerly direction?
☐	☐	☐	46. Is the slide free of any opening between the entrance platform and the sliding surface?
☐	☐	☐	47. Does the slide have a smooth and continuous surface that is free of any gaps or spaces?
☐	☐	☐	48. Is all hardware present, securely attached, and free of significant wear? ▷ *Wear is indicated by visible elongation, deformation, indentation, rust, corrosion, or stripping.*
☐	☐	☐	49. Do bolt ends extend no more than two threads beyond the face of the nut?
☐	☐	☐	50. Are all fastening devices closed to prevent entanglement (see Definitions)?
☐	☐	☐	51. Are wood materials free of warping, wood rot, insect damage, cupping, and checking?
☐	☐	☐	52. Are wood materials free of splinters, heart center, and loose or missing knots?
☐	☐	☐	53. Are metal materials free of rust, corrosion, peeling paint, and bent parts?
☐	☐	☐	54. Are plastic parts unbroken, unchipped, and uncracked, particularly at joints and connections?
☐	☐	☐	55. Is the slide free of chipped, peeling, or worn paint?

16 SPRING ROCKING EQUIPMENT

1 OF 4

PARK NAME   DATE OF INSPECTION   INSPECTOR

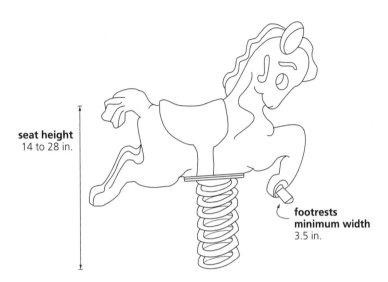

seat height 14 to 28 in.

footrests minimum width 3.5 in.

Yes	No	N/A	AUDIT
☐	☐	☐	1. Does the equipment meet all standards for structural integrity as specified by ASTM F 1487?
☐	☐	☐	2. Does the equipment have an unobstructed use zone? ▷ *The use zones of two spring rockers intended for sitting may overlap. A minimum 72-inch (1800 mm) use zone is required for spring rockers intended for sitting; rockers intended for standing require an 84-inch (2100 mm) use zone that cannot overlap with the use zone of other equipment.*
☐	☐	☐	3. Are vertical angles greater than 55 degrees? ▷ *Inverted angles or angles with a filled apex are exempt (see Definitions).*
☐	☐	☐	4. Is the seat height between 14 and 28 inches (360 and 710 mm) above the safety surface when unloaded and at rest?
☐	☐	☐	5. Is the equipment free of extra holes that could harbor nesting insects? ▷ *This question is based on the authors' opinion and is not addressed by CPSC or ASTM.*

©1997 MIG Communications

16 SPRING ROCKING EQUIPMENT

PARK NAME   DATE OF INSPECTION   INSPECTOR

Yes	No	N/A	AUDIT (cont.)
☐	☐	☐	6. Is the equipment free of pinch, crush, and shear points when operated by a 120-pound (54 kg) user? ▷ *The attachment area of heavy-duty coil springs to the body of the rocking equipment is exempt from these requirements.*
☐	☐	☐	7. Is the equipment free of cables, wires, or other suspended hazards hung within 45 degrees of horizontal (see Definitions)?
☐	☐	☐	8. a. Are handgrips provided for each user?
☐	☐	☐	b. For handgrips intended to be gripped by one hand, is the handgrip at least 3 inches (76 mm) long?
☐	☐	☐	c. For handgrips intended to be gripped by both hands, is the handgrip at least 6 inches (150 mm) long?
☐	☐	☐	9. Are footrests with a minimum width of 3.5 inches (89 mm) provided for each user?
☐	☐	☐	10. Does the seat design minimize the likelihood of use by more than the intended number of users?
☐	☐	☐	11. Are lock washers, self-locking nuts, or other locking means provided for all nuts and bolts to protect them from detachment?
☐	☐	☐	12. Do all metal edges have rolled edging or rounded capping?
☐	☐	☐	13. Are metal materials painted, galvanized, anodized, or composed of non-rusting materials?
☐	☐	☐	14. When located in direct sunlight, have metal materials been coated in plastic to avoid the risk of a contact-burn injury? ▷ *Bare or painted metal surfaces should be avoided in intense, direct sunlight.*
☐	☐	☐	15. Are plastic materials ultraviolet-stabilized to resist fading? ▷ *This question is based on the authors' opinion and is not addressed by CPSC or ASTM.*

TO CONTINUE AUDIT, COMPLETE ANNUAL AND PERIODIC INSPECTIONS.

16 SPRING ROCKING EQUIPMENT

PARK NAME   DATE OF INSPECTION   INSPECTOR

Yes	No	N/A		ANNUAL INSPECTION
☐	☐	☐	16.	Is the equipment free of head and neck entrapments (see Inspection Procedures)?
☐	☐	☐	17.	Do protrusions meet the protrusion test criteria (see Inspection Procedures)?
☐	☐	☐	18.	Is the equipment free of hollow support posts or tubes with open ends?
☐	☐	☐	19.	Are equipment footings securely anchored?
☐	☐	☐	20.	Are wood materials naturally rot- and insect-resistant, or treated with a wood preservative below and up to 6 inches (150 mm) above the surface of the play area?
				If a wood preservative was used, list the preservative's name: _____.
☐	☐	☐	21.	Is the wood preservative safe for use in children's play areas, as specified by ASTM F 1487 standards?
☐	☐	☐	22.	Are paints free of lead (0.06% maximum lead by dry weight) as specified by ASTM F 1487 standards?

TO CONTINUE ANNUAL INSPECTION, COMPLETE PERIODIC INSPECTION.

16 SPRING ROCKING EQUIPMENT

PARK NAME / DATE OF INSPECTION / INSPECTOR

Yes	No	N/A	**PERIODIC INSPECTION**
☐	☐	☐	23. Is the equipment stable and without severe structural deterioration, such as at the footings and joints?
☐	☐	☐	24. Is the equipment free from loose, missing, or broken parts and vandalism?
☐	☐	☐	25. Is the equipment free of sharp points, corners, or edges?
☐	☐	☐	26. Is all hardware present, securely attached, and free of significant wear? ▷ *Wear is indicated by visible elongation, deformation, indentation, rust, corrosion, or stripping.*
☐	☐	☐	27. Do bolt ends extend no more than two threads beyond the face of the nut?
☐	☐	☐	28. Are all fastening devices closed to prevent entanglement (see Definitions)?
☐	☐	☐	29. Are wood materials free of warping, wood rot, insect damage, cupping, and checking?
☐	☐	☐	30. Are wood materials free of splinters, heart center, and loose or missing knots?
☐	☐	☐	31. Are metal materials free of rust, corrosion, peeling paint, and bent parts?
☐	☐	☐	32. Are plastic parts unbroken, unchipped, and uncracked, particularly at joints and connections?
☐	☐	☐	33. Is the equipment free of chipped, peeling, or worn paint?

©1997 MIG Communications

17 SWINGS

1 OF 5

PARK NAME   **DATE OF INSPECTION**   **INSPECTOR**

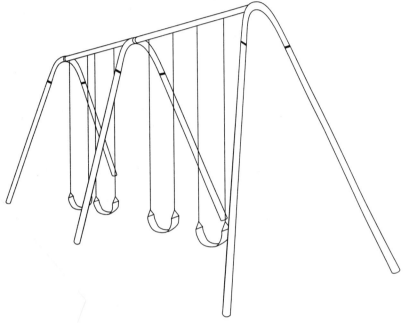

AUDIT

☐   1.   Does the swing meet all standards for structural integrity as specified by ASTM F 1487?

☐   2.   Is the swing use zone free of any obstruction (see Definitions)?
▷ *For swings with belt seats, the length of the swing use zone is equal to two times the distance from the safety surface to the swing pivot point. The use zone should extend for this distance to both the front and rear of the crossbeam for a width at least as wide as the beam. For swings with enclosed seats, such as tot swings or bucket swings, the distance provided to the front and rear of the crossbeam should be equal to twice the distance measured from the top of the occupant's sitting surface to the swing pivot point. For both belt swings and swings with enclosed seats, a 72-inch (1800 mm) use zone should extend out from both sides of the swing support. When swings are located adjacent to each other, the swings may share the 72-inch (1800 mm) use zone at the side.*

☐   3.   Are vertical angles greater than 55 degrees?
▷ *Inverted angles or angles with a filled apex are exempt (see Definitions).*

17 SWINGS

| PARK NAME | DATE OF INSPECTION | INSPECTOR |

Yes	No	N/A	AUDIT (cont.)
☐	☐	☐	4. a. Are tot swing seats suspended at least 24 inches (600 mm) above the safety surface? ▷ *This question is based on the authors' opinion and is not addressed by CPSC or ASTM.*
☐	☐	☐	b. Are swing seats suspended at least 12 inches (300 mm) above the safety surface? ▷ *Seats should not be suspended at a height less than 12 inches (300 mm) when occupied by the maximum user.*
☐	☐	☐	5. Is the swing free of extra holes that could harbor nesting insects? ▷ *This question is based on the authors' opinion and is not addressed by CPSC or ASTM.*
☐	☐	☐	6. Is the swing free of pinch, crush, and shear points (see Definitions)? ▷ *Swing chains are exempt.*
☐	☐	☐	7. Is the swing free of cables, wires, or other suspended hazards hung within 45 degrees of horizontal (see Definitions)?
☐	☐	☐	8. Does the swing support frame design discourage climbing?
☐	☐	☐	9. Does the swing support frame not include a designated play surface?
☐	☐	☐	10. Are the swings located on the outer edge of the play area, away from other play equipment and activity areas?
☐	☐	☐	11. Are the swings freestanding and not attached to a composite structure?
☐	☐	☐	12. Is a maximum of two swings hung in each section of the support structure?
☐	☐	☐	13. Are the swings free of hard or heavy seats?
☐	☐	☐	14. Do swing seats accommodate only one user?
☐	☐	☐	15. a. Are tot swings hung from a separate support structure or swing bay section? ▷ *It is recommended that tot swings and belt swings be hung from separate structures or separate bay sections.*
☐	☐	☐	b. Do tot swings provide 360-degree support?

©1997 MIG Communications

17 SWINGS

PARK NAME   DATE OF INSPECTION   INSPECTOR

Yes	No	N/A	AUDIT (cont.)
☐	☐	☐	16. Are swing seats spaced at least 24 inches (600 mm) apart when occupied by the maximum user? ▷ *Measure the distance at a height of 60 inches (1500 mm) above the safety surface.*
☐	☐	☐	17. Are swing seats spaced at least 30 inches (760 mm) from the swing support structure when occupied by the maximum user? ▷ *Measure the distance at a height of 60 inches (1500 mm) above the safety surface.*
☐	☐	☐	18. Are swing hangers spaced wider than the width of the swing seat to reduce side-to-side motion? ▷ *Swing hangers are the hardware from which the swing chains are suspended.*
☐	☐	☐	19. Is the distance between swing hangers supporting one swing seat at least 20 inches (510 mm) apart and greater than the width of the seat when occupied by the maximum user?
☐	☐	☐	20. Do chains or cables meet ASTM F 1487 structural integrity requirements?
☐	☐	☐	21. Do cables measure at least 1 inch (25 mm) in diameter?
☐	☐	☐	22. Are lock washers, self-locking nuts, or other locking means provided for all nuts and bolts to protect them from detachment?
☐	☐	☐	23. Do all metal edges have rolled edging or rounded capping?
☐	☐	☐	24. Are metal materials painted, galvanized, anodized, or composed of non-rusting materials?
☐	☐	☐	25. When located in direct sunlight, have metal materials been coated in plastic to avoid the risk of a contact-burn injury? ▷ *Bare or painted metal surfaces should be avoided in intense, direct sunlight.*
☐	☐	☐	26. Are plastic materials ultraviolet-stabilized to resist fading? ▷ *This question is based on the authors' opinion and is not addressed by CPSC or ASTM.*

TO CONTINUE AUDIT, COMPLETE ANNUAL AND PERIODIC INSPECTIONS.

17 SWINGS

PARK NAME DATE OF INSPECTION INSPECTOR

Yes	No	N/A	**ANNUAL INSPECTION**
☐	☐	☐	27. Is the swing free of head and neck entrapments (see Inspection Procedures)?
☐	☐	☐	28. Do protrusions meet the protrusion test criteria (see Inspection Procedures)?
☐	☐	☐	29. Is the swing free of hollow support posts or tubes with open ends?
☐	☐	☐	30. Are equipment footings securely anchored?
☐	☐	☐	31. Are wood materials naturally rot- and insect-resistant, or treated with a wood preservative below and up to 6 inches (150 mm) above the surface of the play area?

If a wood preservative was used, list the preservative's name:

_____.

Yes	No	N/A	
☐	☐	☐	32. Is the wood preservative safe for use in children's play areas, as specified by ASTM F 1487 standards?
☐	☐	☐	33. Are paints free of lead (0.06% maximum lead by dry weight) as specified by ASTM F 1487 standards?

TO CONTINUE ANNUAL INSPECTION, COMPLETE PERIODIC INSPECTION.

17 SWINGS

PARK NAME DATE OF INSPECTION INSPECTOR

Yes	No	N/A	**PERIODIC INSPECTION**
☐	☐	☐	34. Is the swing stable and without severe structural deterioration, such as at the footings and joints?
☐	☐	☐	35. Is the swing free of loose, missing, or broken parts and vandalism?
☐	☐	☐	36. Is the swing free of sharp points, corners, or edges?
☐	☐	☐	37. For metal swings and swing chains, is the outdoor temperature above freezing when in use?
☐	☐	☐	38. Are chains without significant wear? ▷ *Wear is indicated by visible elongation, deformation, indentation, rust, or corrosion.*
☐	☐	☐	39. Are cables free of frayed or projecting wires?
☐	☐	☐	40. Is all hardware present, securely attached, and free of significant wear? ▷ *Wear is indicated by visible elongation, deformation, indentation, rust, corrosion, or stripping.*
☐	☐	☐	41. Do bolt ends extend no more than two threads beyond the face of the nut?
☐	☐	☐	42. Are all fastening devices closed to prevent entanglement (see Definitions)?
☐	☐	☐	43. Are all swing chains or cables connected to the crossbeam with bearings that reduce friction and wear? ▷ *A steel cable permanently affixed to a hanger assembly meets this requirement.*
☐	☐	☐	44. Are swing bearings in good condition and well lubricated?
☐	☐	☐	45. Are wood materials free of warping, wood rot, insect damage, cupping, and checking?
☐	☐	☐	46. Are wood materials free of splinters, heart center, and loose or missing knots?
☐	☐	☐	47. Are metal materials free of rust, corrosion, peeling paint, and bent parts?
☐	☐	☐	48. Are plastic parts unbroken, unchipped, and uncracked, particularly at joints and connections?
☐	☐	☐	49. Is the swing free of chipped, peeling, or worn paint?

18 SWINGS, ROTATING

PARK NAME _____ DATE OF INSPECTION _____ INSPECTOR _____

Note: In the authors' opinion, rotating swings are not recommended for children under 3.

support structure

minimum height 12 in.

Yes	No	N/A	AUDIT
☐	☐	☐	1. Does the swing meet all standards for structural integrity as specified by ASTM F 1487?
☐	☐	☐	2. Does the swing have a minimum 72-inch (1800 mm) unobstructed use zone extending in all directions from the swing support structure (see Definitions)? ▷ *Adjacent swing support structures may share the 72-inch (1800 mm) use zone on the side.*
☐	☐	☐	3. Does the swing have a use zone that extends in all directions from the swing seat and equals the vertical distance between the pivot point and the swing seat plus 72 inches (1800 mm) (see Definitions)?
☐	☐	☐	4. Are vertical angles greater than 55 degrees? ▷ *Inverted angles or angles with a filled apex are exempt (see Definitions).*
☐	☐	☐	5. Is the lower edge of the rotating swing at least 12 inches (300 mm) above the playing surface when occupied by the maximum user?
☐	☐	☐	6. Is the swing free of extra holes that could harbor nesting insects? ▷ *This question is based on the authors' opinion and is not addressed by CPSC or ASTM.*
☐	☐	☐	7. Is the swing free of pinch, crush, and shear points (see Definitions)? ▷ *Swing chains are exempt.*

©1997 MIG Communications

18 SWINGS, ROTATING

PARK NAME _____ DATE OF INSPECTION _____ INSPECTOR _____

Yes	No	N/A	AUDIT (cont.)
☐	☐	☐	8. Is the swing free of cables, wires, or other suspended hazards hung within 45 degrees of horizontal (see Definitions)?
☐	☐	☐	9. Is the swing located at the edge of the play area, away from other play equipment and activity areas?
☐	☐	☐	10. Is the swing freestanding and not attached to a composite structure?
☐	☐	☐	11. Does the swing support structure discourage climbing?
☐	☐	☐	12. Is a maximum of one rotating swing hung in one swing bay?
☐	☐	☐	13. Are rotating swings hung from a separate support structure than to-fro swings?
☐	☐	☐	14. Is the swing seat free of openings that could harbor insects or trap water?
☐	☐	☐	15. Is the swing seat smooth, having rounded edges with a minimum curvature radius of ¼ inch (6 mm)? ▷ *Steel-belted automobile tires, hard seats, or seats that exceed 35 pounds (15.8 kg) when unoccupied should not be used.*
☐	☐	☐	16. Does the swing have an unobstructed clearance zone (see Definitions)?
☐	☐	☐	17. Do chains or cables meet ASTM F 1487 structural integrity requirements?
☐	☐	☐	18. Are chains or cables without significant wear? ▷ *Wear is indicated by visible elongation, deformation, indentation, rust, or corrosion.*
☐	☐	☐	19. Do cables measure at least 1 inch (25 mm) in diameter?
☐	☐	☐	20. Are lock washers, self-locking nuts, or other locking means provided for all nuts and bolts to protect them from detachment?
☐	☐	☐	21. Do all metal edges have rolled edging or rounded capping?
☐	☐	☐	22. Are metal materials painted, galvanized, anodized, or composed of non-rusting materials?
☐	☐	☐	23. When located in direct sunlight, have metal materials been coated in plastic to avoid the risk of a contact-burn injury? ▷ *Bare or painted metal surfaces should be avoided in intense, direct sunlight.*
☐	☐	☐	24. Are plastic materials ultraviolet-stabilized to resist fading? ▷ *This question is based on the authors' opinion and is not addressed by CPSC or ASTM.*

TO CONTINUE AUDIT, COMPLETE ANNUAL AND PERIODIC INSPECTIONS.

©1997 MIG Communications

18 SWINGS, ROTATING

PARK NAME DATE OF INSPECTION INSPECTOR

Yes	No	N/A	**ANNUAL INSPECTION**
☐	☐	☐	25. Is the swing free of head and neck entrapments (see Inspection Procedures)?
☐	☐	☐	26. Do protrusions meet the protrusion test criteria (see Inspection Procedures)?
☐	☐	☐	27. Are equipment footings securely anchored?
☐	☐	☐	28. Are wood materials naturally rot- and insect-resistant, or treated with a wood preservative below and up to 6 inches (150 mm) above the surface of the play area?

If a wood preservative was used, list the preservative's name:

_____.

Yes	No	N/A	
☐	☐	☐	29. Is the wood preservative safe for use in children's play areas, as specified by ASTM F 1487 standards?
☐	☐	☐	30. Are paints free of lead (0.06% maximum lead by dry weight) as specified by ASTM F 1487 standards?

TO CONTINUE ANNUAL INSPECTION, COMPLETE PERIODIC INSPECTION.

18 SWINGS, ROTATING

PARK NAME   DATE OF INSPECTION   INSPECTOR

Yes	No	N/A	**PERIODIC INSPECTION**
☐	☐	☐	31. Is the swing stable and without severe structural deterioration, such as at the footings and joints?
☐	☐	☐	32. Is the swing free of loose, missing, or broken parts and vandalism?
☐	☐	☐	33. Is the swing free of sharp points, corners, or edges?
☐	☐	☐	34. Are cables free of frayed or projecting wires?
☐	☐	☐	35. Is all hardware present, securely attached, and free of significant wear? ▷ *Wear is indicated by visible elongation, deformation, indentation, rust, corrosion, or stripping.*
☐	☐	☐	36. Do bolt ends extend no more than two threads beyond the face of the nut?
☐	☐	☐	37. Are all fastening devices closed to prevent entanglement (see Definitions)?
☐	☐	☐	38. Does the swing hanger have bearings, bushings, or other means of reducing friction and wear at the pivot point?
☐	☐	☐	39. Are bearings in good condition and well lubricated?
☐	☐	☐	40. Are wood materials free of warping, wood rot, insect damage, cupping, and checking?
☐	☐	☐	41. Are wood materials free of splinters, heart center, and loose or missing knots?
☐	☐	☐	42. Are metal materials free of rust, corrosion, peeling paint, and bent parts?
☐	☐	☐	43. Are plastic parts unbroken, unchipped, and uncracked, particularly at joints and connections?
☐	☐	☐	44. Is the swing free of chipped, peeling, or worn paint?

©1997 MIG Communications

19 TRACK RIDES

PARK NAME　　　　**DATE OF INSPECTION**　　　　**INSPECTOR**

Note: According to ASTM F 1487, track rides are not recommended for children under 5.

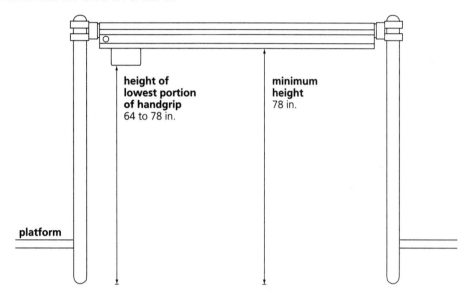

Yes	No	N/A	**AUDIT**
☐	☐	☐	1. Does the track ride meet all standards for structural integrity as specified by ASTM F 1487?
☐	☐	☐	2. Does the track ride have a 72-inch (1800 mm) unobstructed use zone?
☐	☐	☐	3. Are vertical angles greater than 55 degrees? ▷ *Inverted angles or angles with a filled apex are exempt (see Definitions).*
☐	☐	☐	4. For 5- to 12-year-olds, is the track ride at least 78 inches (1950 mm) high?
☐	☐	☐	5. Is the track ride free of extra holes that could harbor nesting insects? ▷ *This question is based on the authors' opinion and is not addressed by CPSC or ASTM.*
☐	☐	☐	6. Is the track ride free of pinch, crush, and shear points (see Definitions)? ▷ *When the rolling portions of the handgrip are enclosed within the track beam, the track assembly is exempt from pinch, crush, and shear requirements.*

19 TRACK RIDES

PARK NAME   **DATE OF INSPECTION**   **INSPECTOR**

Yes	No	N/A	AUDIT (cont.)
☐	☐	☐	7. Is the track ride free of cables, wires, or other suspended hazards hung within 45 degrees of horizontal (see Definitions)?
☐	☐	☐	8. Is the handgrip or first ring inset so that it is not directly located above the platform or rungs used for equipment entry or exit?
☐	☐	☐	9. Do the handgrips measure between 0.95 and 1.55 inches (24.1 and 39.4 mm) in diameter?
			10. For track rides, does the handgripping component meet the following criteria:
☐	☐	☐	a. Is the lowest portion of the handgrip 64 to 78 inches (1600 to 1950 mm) above the safety surface?
☐	☐	☐	b. Are all parts of the handgrip not removable with or without the use of tools unless the handgrip is removed from the beam?
☐	☐	☐	11. Is the horizontal distance to the first handhold no more than 10 inches (250 mm) from the leading edge of the takeoff and landing platforms?
☐	☐	☐	12. Is the distance from the takeoff and landing structures to the nearest handhold no more than 45 inches (1140 mm)?
☐	☐	☐	13. Do chains or cables meet ASTM F 1487 structural integrity requirements?
☐	☐	☐	14. Do cables measure at least 1 inch (25 mm) in diameter?
☐	☐	☐	15. Do all metal edges have rolled edging or rounded capping?
☐	☐	☐	16. Are metal materials painted, galvanized, anodized, or composed of non-rusting materials?
☐	☐	☐	17. When located in direct sunlight, have metal materials been coated in plastic to avoid the risk of a contact-burn injury? ▷ *Bare or painted metal surfaces should be avoided in intense, direct sunlight.*
☐	☐	☐	18. Are plastic materials ultraviolet-stabilized to resist fading? ▷ *This question is based on the authors' opinion and is not addressed by CPSC or ASTM.*

TO CONTINUE AUDIT, COMPLETE ANNUAL AND PERIODIC INSPECTIONS.

©1997 MIG Communications

19 TRACK RIDES

PARK NAME DATE OF INSPECTION INSPECTOR

Yes	No	N/A	**ANNUAL INSPECTION**
☐	☐	☐	19. Is the track ride free of head and neck entrapments (see Inspection Procedures)?
☐	☐	☐	20. Do protrusions meet the protrusion test criteria (see Inspection Procedures)?
☐	☐	☐	21. Is the track ride free of hollow support posts or tubes with open ends?
☐	☐	☐	22. Are equipment footings securely anchored?
☐	☐	☐	23. Are wood materials naturally rot- and insect-resistant, or treated with a wood preservative below and up to 6 inches (150 mm) above the surface of the play area?

If a wood preservative was used, list the preservative's name:

_____.

Yes	No	N/A	
☐	☐	☐	24. Is the wood preservative safe for use in children's play areas, as specified by ASTM F 1487 standards?
☐	☐	☐	25. Are paints free of lead (0.06% maximum lead by dry weight) as specified by ASTM F 1487 standards?

TO CONTINUE ANNUAL INSPECTION, COMPLETE PERIODIC INSPECTION.

19 TRACK RIDES

PARK NAME   DATE OF INSPECTION   INSPECTOR

Yes	No	N/A	**PERIODIC INSPECTION**
☐	☐	☐	26. Is the track ride stable and without severe structural deterioration, such as at the footings and joints?
☐	☐	☐	27. Is the track ride free of loose, missing, or broken parts and vandalism?
☐	☐	☐	28. Is the track ride free of sharp points, corners, or edges?
☐	☐	☐	29. Are the takeoff and landing areas free from obstructions?
☐	☐	☐	30. Are chains or cables without significant wear? ▷ *Wear is indicated by visible elongation, deformation, indentation, rust, or corrosion.*
☐	☐	☐	31. Are cables free of frayed or projecting wires?
☐	☐	☐	32. Is all hardware present, securely attached, and free of significant wear? ▷ *Wear is indicated by visible elongation, deformation, indentation, rust, corrosion, or stripping.*
☐	☐	☐	33. Do bolt ends extend no more than two threads beyond the face of the nut?
☐	☐	☐	34. Are lock washers, self-locking nuts, or other locking means provided for all nuts and bolts to protect them from detachment?
☐	☐	☐	35. Are all fastening devices closed to prevent entanglement (see Definitions)?
☐	☐	☐	36. Are all moving suspended elements connected to the fixed support with bearings that reduce friction and wear? ▷ *A steel cable permanently affixed to a hanger assembly meets this requirement.*
☐	☐	☐	37. Are the bearings in good condition and well lubricated?
☐	☐	☐	38. Are wood materials free of warping, wood rot, insect damage, cupping, and checking?
☐	☐	☐	39. Are wood materials free of splinters, heart center, and loose or missing knots?
☐	☐	☐	40. Are metal materials free of rust, corrosion, peeling paint, and bent parts?
☐	☐	☐	41. Are plastic parts unbroken, unchipped, and uncracked, particularly at joints and connections?
☐	☐	☐	42. Is the track ride free of chipped, peeling, or worn paint?

©1997 MIG Communications

20 TUNNELS

PARK NAME _____ DATE OF INSPECTION _____ INSPECTOR _____

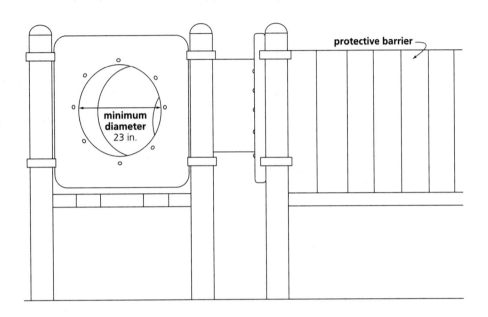

Yes	No	N/A	AUDIT
☐	☐	☐	1. Does the tunnel meet all standards for structural integrity as specified by ASTM F 1487?
☐	☐	☐	2. Does the tunnel have a 72-inch (1800 mm) unobstructed use zone?
☐	☐	☐	3. Are vertical angles greater than 55 degrees? ▷ *Inverted angles or angles with a filled apex are exempt (see Definitions).*
☐	☐	☐	4. Is the interior diameter of the tunnel at least 23 inches (580 mm)?
☐	☐	☐	5. Is the tunnel free of extra holes that could harbor nesting insects? ▷ *This question is based on the authors' opinion and is not addressed by CPSC or ASTM.*
☐	☐	☐	6. Is the tunnel free of pinch, crush, and shear points (see Definitions)?
☐	☐	☐	7. Are all tunnel edges rounded?
☐	☐	☐	8. Is the tunnel free of cables, wires, or other suspended hazards hung within 45 degrees of horizontal (see Definitions)?

©1997 MIG Communications

20 TUNNELS

PARK NAME   DATE OF INSPECTION   INSPECTOR

Yes	No	N/A	AUDIT (cont.)
☐	☐	☐	9. Do all metal edges have rolled edging or rounded capping?
☐	☐	☐	10. Are metal materials painted, galvanized, anodized, or composed of non-rusting materials?
☐	☐	☐	11. When located in direct sunlight, have metal materials been coated in plastic to avoid the risk of a contact-burn injury? ▷ *Bare or painted metal surfaces should be avoided in intense, direct sunlight.*
☐	☐	☐	12. Are plastic materials ultraviolet-stabilized to resist fading? ▷ *This question is based on the authors' opinion and is not addressed by CPSC or ASTM.*

TO CONTINUE AUDIT, COMPLETE ANNUAL AND PERIODIC INSPECTIONS.

20 TUNNELS

PARK NAME   **DATE OF INSPECTION**   **INSPECTOR**

Yes	No	N/A		ANNUAL INSPECTION
☐	☐	☐	13.	Is the tunnel free of head and neck entrapments (see Inspection Procedures)?
☐	☐	☐	14.	Do protrusions meet the protrusion test criteria (see Inspection Procedures)?
☐	☐	☐	15.	Is the tunnel free of hollow support posts or tubes with open ends?
☐	☐	☐	16.	Are equipment footings securely anchored?
☐	☐	☐	17.	Are wood materials naturally rot- and insect-resistant, or treated with a wood preservative below and up to 6 inches (150 mm) above the surface of the play area? If a wood preservative was used, list the preservative's name: _____.
☐	☐	☐	18.	Is the wood preservative safe for use in children's play areas, as specified by ASTM F 1487 standards?
☐	☐	☐	19.	Are paints free of lead (0.06% maximum lead by dry weight) as specified by ASTM F 1487 standards?

TO CONTINUE ANNUAL INSPECTION, COMPLETE PERIODIC INSPECTION.

20 TUNNELS

PARK NAME _____ DATE OF INSPECTION _____ INSPECTOR _____

Yes	No	N/A	**PERIODIC INSPECTION**
☐	☐	☐	20. Is the tunnel stable and without severe structural deterioration, such as at the footings and joints?
☐	☐	☐	21. Is the tunnel free of loose, missing, or broken parts and vandalism?
☐	☐	☐	22. Is the tunnel free of sharp points, corners, or edges?
☐	☐	☐	23. Is all hardware present, securely attached, and free of significant wear? ▷ *Wear is indicated by visible elongation, deformation, indentation, rust, corrosion, or stripping.*
☐	☐	☐	24. Do bolt ends extend no more than two threads beyond the face of the nut?
☐	☐	☐	25. Are lock washers, self-locking nuts, or other locking means provided for all nuts and bolts to protect them from detachment?
☐	☐	☐	26. Are all fastening devices closed to prevent entanglement (see Definitions)?
☐	☐	☐	27. Are wood materials free of warping, wood rot, insect damage, cupping, and checking?
☐	☐	☐	28. Are wood materials free of splinters, heart center, and loose or missing knots?
☐	☐	☐	29. Are metal materials free of rust, corrosion, peeling paint, and bent parts?
☐	☐	☐	30. Are plastic parts unbroken, unchipped, and uncracked, particularly at joints and connections?
☐	☐	☐	31. Is the tunnel free of chipped, peeling, or worn paint?

©1997 MIG Communications

21 COMPOSITE STRUCTURES

PARK NAME   DATE OF INSPECTION   INSPECTOR

PLAY EVENTS

▷ *To inspect a composite structure, use this checklist plus the separate checklists for each play event attached to the structure.*

- ☐ ladders
- ☐ stairways
- ☐ ramps
- ☐ guardrails & protective barriers
- ☐ balance beams
- ☐ bars, chin-up & turning
- ☐ bars, parallel
- ☐ bridges, clatter
- ☐ bridges, stationary
- ☐ climbers
- ☐ climbers, flexible
- ☐ fire poles
- ☐ horizontal ladders & ring treks
- ☐ playhouses
- ☐ slides
- ☐ spring rocking equipment
- ☐ swings
- ☐ swings, rotating
- ☐ track rides
- ☐ tunnels
- ☐ other: _____

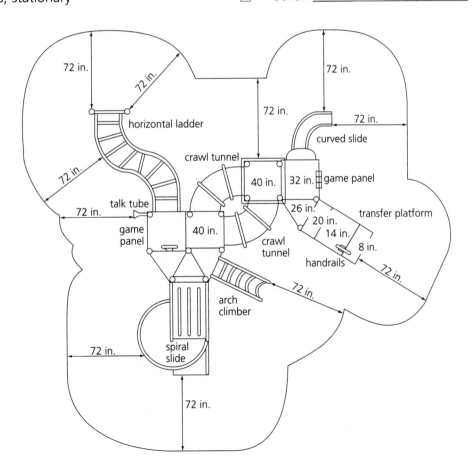

©1997 MIG Communications

21 COMPOSITE STRUCTURES

PARK NAME _____ DATE OF INSPECTION _____ INSPECTOR _____

Yes	No	N/A		AUDIT
☐	☐	☐	1.	Does the composite structure meet all standards for structural integrity as specified by ASTM F 1487?
☐	☐	☐	2.	For 2- to 12-year-olds, is the composite structure's use zone free of obstructions for a distance equal to the use zones recommended for all individual play events (minimum 72 inches [1800 mm])?
☐	☐	☐	3.	Are vertical angles greater than 55 degrees? ▷ *Inverted angles or angles with a filled apex are exempt (see Definitions).*
☐	☐	☐	4.	Is the composite structure free of extra holes that could harbor nesting insects? ▷ *This question is based on the authors' opinion and is not addressed by CPSC or ASTM.*
☐	☐	☐	5.	Is the composite structure free of pinch, crush, and shear points (see Definitions)?
☐	☐	☐	6.	Is the composite structure free of cables, wires, or other suspended hazards hung within 45 degrees of horizontal (see Definitions)?
☐	☐	☐	7.	a. For 2- to 5-year-olds, are all play platforms that are more than 30 inches (760 mm) high enclosed by a protective barrier 29 inches (740 mm) or greater in height?
☐	☐	☐		b. For 5- to 12-year-olds, are all play platforms that are more than 48 inches (1200 mm) high enclosed by a protective barrier 38 inches (970 mm) or greater in height?

▷ *See Guardrails & Protective Barriers checklist for additional inspection items.*

☐	☐	☐	8.	a. For 2- to 5-year-olds, are all play equipment platforms over 20 inches (510 mm) high enclosed by a guardrail that is a maximum 23 inches (580 mm) high at the lower edge and 29 inches (740 mm) high at the top edge?
☐	☐	☐		b. For 5- to 12-year-olds, are all play equipment platforms over 30 inches (760 mm) high enclosed by a guardrail that is a maximum 28 inches (710 mm) high at the lower edge and 38 inches (970 mm) high at the top edge?
☐	☐	☐	9.	Is the composite structure free of play events or components in its interior onto which a child may fall from a height greater than 18 inches (450 mm)?

©1997 MIG Communications

21 COMPOSITE STRUCTURES

PARK NAME   DATE OF INSPECTION   INSPECTOR

Yes	No	N/A	AUDIT (cont.)
☐	☐	☐	10. Are handrails or handgrips provided to ease the transition between platforms and attached play events?
☐	☐	☐	11. a. For 2- to 5-year-olds, do adjacent platforms that have a height difference greater than 12 inches (300 mm) have a handgrip or handrail to ease the transition between platforms?
☐	☐	☐	b. For 5- to 12-year-olds, do adjacent platforms that have a height difference greater than 18 inches (460 mm) have a handgrip or handrail to ease the transition between platforms? ▷ *See Equipment Access & Egress checklist for handrail and handgrip requirements.*
☐	☐	☐	12. Are the platforms level (within 2 degrees of the horizontal)?
☐	☐	☐	13. Are openings provided in the platforms to allow for drainage?
☐	☐	☐	14. For 2- to 5-year-olds, is there another means of equipment access (e.g., ramp, stairway, or stepladder) in addition to a climbing apparatus?
☐	☐	☐	15. Do chains or cables meet ASTM F 1487 structural integrity requirements?
☐	☐	☐	16. Do cables measure at least 1 inch (25 mm) in diameter?
☐	☐	☐	17. Are lock washers, self-locking nuts, or other locking means provided for all nuts and bolts to protect them from detachment?
☐	☐	☐	18. Do all metal edges have rolled edging or rounded capping?
☐	☐	☐	19. Are metal materials painted, galvanized, anodized, or composed of non-rusting materials?
☐	☐	☐	20. When located in direct sunlight, have metal materials been coated in plastic to avoid the risk of a contact-burn injury? ▷ *Bare or painted metal surfaces should be avoided in intense, direct sunlight.*
☐	☐	☐	21. Are plastic materials ultraviolet-stabilized to resist fading? ▷ *This question is based on the authors' opinion and is not addressed by CPSC or ASTM.*

TO CONTINUE AUDIT, COMPLETE ANNUAL AND PERIODIC INSPECTIONS.

©1997 MIG Communications

21 COMPOSITE STRUCTURES

PARK NAME   DATE OF INSPECTION   INSPECTOR

Yes	No	N/A	**ANNUAL INSPECTION**
☐	☐	☐	22. Is the composite structure free of head and neck entrapments (see Inspection Procedures)?
☐	☐	☐	23. Do protrusions meet the protrusion test criteria (see Inspection Procedures)?
☐	☐	☐	24. Is the composite structure free of hollow support posts or tubes with open ends?
☐	☐	☐	25. Are equipment footings securely anchored?
☐	☐	☐	26. Are wood materials naturally rot- and insect-resistant, or treated with a wood preservative below and up to 6 inches (150 mm) above the surface of the play area?

If a wood preservative was used, list the preservative's name:

_____.

Yes	No	N/A	
☐	☐	☐	27. Is the wood preservative safe for use in children's play areas, as specified by ASTM F 1487 standards?
☐	☐	☐	28. Are paints free of lead (0.06% maximum lead by dry weight) as specified by ASTM F 1487 standards?

TO CONTINUE ANNUAL INSPECTION, COMPLETE PERIODIC INSPECTION.

©1997 MIG Communications

21 COMPOSITE STRUCTURES

PARK NAME　　　DATE OF INSPECTION　　　INSPECTOR

Yes	No	N/A	**PERIODIC INSPECTION**
☐	☐	☐	29. Is the composite structure stable and without severe structural deterioration, such as at the footings and joints?
☐	☐	☐	30. Is the composite structure free of loose, missing, or broken parts and vandalism?
☐	☐	☐	31. Is the composite structure free of wet or icy surfaces?
☐	☐	☐	32. Is the composite structure free of sharp points, corners, or edges?
☐	☐	☐	33. Are chains or cables without significant wear? ▷ *Wear is indicated by visible elongation, deformation, indentation, rust, or corrosion.*
☐	☐	☐	34. Are cables free of frayed or projecting wires?
☐	☐	☐	35. Are cables or chains fixed tightly at both ends so that there is no possibility of overlapping and entrapping a child? ▷ *Swing chains are exempt from this requirement.*
☐	☐	☐	36. Is all hardware present, securely attached, and free of significant wear? ▷ *Wear is indicated by visible elongation, deformation, indentation, rust, corrosion, or stripping.*
☐	☐	☐	37. Do bolt ends extend no more than two threads beyond the face of the nut?
☐	☐	☐	38. Are all fastening devices closed to prevent entanglement (see Definitions)?
☐	☐	☐	39. Are all moving suspended elements connected to the fixed support with bearings that reduce friction and wear? ▷ *A steel cable permanently affixed to a hanger assembly meets this requirement.*
☐	☐	☐	40. Are the bearings in good condition and well lubricated?
☐	☐	☐	41. Are wood materials free of warping, wood rot, insect damage, cupping, and checking?
☐	☐	☐	42. Are wood materials free of splinters, heart center, and loose or missing knots?
☐	☐	☐	43. Are metal materials free of rust, corrosion, peeling paint, and bent parts?

©1997 MIG Communications

21 COMPOSITE STRUCTURES

PARK NAME

DATE OF INSPECTION

INSPECTOR

Yes	No	N/A	**PERIODIC INSPECTION (cont.)**
☐	☐	☐	44. Are plastic parts unbroken, unchipped, and uncracked, particularly at joints and connections?
☐	☐	☐	45. Is the composite structure free of chipped, peeling, or worn paint?

SAFETY INSPECTION SUMMARY

AGENCY NAME	PARK NAME	
TYPE OF INSPECTION ☐ AUDIT ☐ ANNUAL ☐ PERIODIC ☐ DAILY	DATE OF INSPECTION	TOTAL INSPECTION TIME (MIN.)
INSPECTOR(S)		

INSPECTION SUMMARY
☐ NO HAZARDS FOUND   ☐ THE FOLLOWING HAZARDS WERE NOTED AND FOLLOW-UP ACTIONS TAKEN:

HAZARDS IDENTIFIED (DESCRIBE HAZARD & LOCATION)	CORRECTIVE ACTIONS		
	MAINTENANCE TASK PERFORMED (DESCRIBE)	AREA CLOSED (ENTER DATE)	WORK ORDER SUBMITTED (ENTER DATE)

I VERIFY THAT I HAVE COMPLETED THE ABOVE INSPECTION AND NOTED ALL HAZARDS AND FOLLOW-UP ACTIONS.

SIGNATURE OF INSPECTOR(S)	SIGNATURE OF SUPERVISOR

©1997 MIG Communications

BIBLIOGRAPHY

American Society for Testing and Materials (1993, 1995). F 1487. *Standard Consumer Safety Performance Specification for Playground Equipment for Public Use.* West Conshohocken, Pa.: ASTM.

——— (1993). F 1292. *Standard Specification for Impact Attenuation of Surface Systems Under and Around Playground Equipment.* West Conshohocken, Pa.: ASTM.

Bruya, L.D. (1988). *Play Spaces for Children: A New Beginning.* Vol. 2. Reston, Va.: American Alliance for Health, Physical Education, Recreation, and Dance.

Bruya, L.D., and S.J. Langendorfer (1988). *Where Our Children Play: Elementary School Playground Equipment.* Vol. 1. Reston, Va.: American Alliance for Health, Physical Education, Recreation, and Dance.

Christiansen, M.L., ed. (1993). *Points about Playgrounds.* Arlington, Va.: National Recreation and Park Association.

——— (1992). *Play It Safe: An Anthology of Playground Safety.* Arlington, Va.: National Recreation and Park Association.

Freedberg, L. (1983). *America's Poisoned Playgrounds: Children and Toxic Chemicals.* Oakland, Calif.: Youth News.

Gold, S.M. (1981). Designing Public Playgrounds for User Safety. *Australian Parks and Recreation* 22, no. 3: 10–14.

——— (1988). Playground Design: The Standard of Care. *California Council of Landscape Architects Quarterly.*

——— (1988). Safety Checklists for Parks and Recreation Areas. Proceedings of the California Park and Recreation Society Conference, Long Beach, Calif.

Goltsman, S.M., T.A. Gilbert, and S.D. Wohlford (1993). *The Accessibility Checklist: An Evaluation System for Buildings and Outdoor Settings.* Berkeley, Calif.: MIG Communications.

Kutska, K.S., and K.J. Hoffman (1992). *Playground Safety Is No Accident.* Arlington, Va.: National Recreation and Park Association.

Landscape Structures, Inc. (1989). *Playground Maintenance.* Delano, Minn.: Landscape Structures, Inc.

MIG, Inc. (1993). *Children's Outdoor Play Areas, Criteria Search and Analysis Report.* Huntsville, Ala.: U.S. Army Corps of Engineers.

Moore, R.C. (1993). *Plants for Play: A Plant Selection Guide for Children's Outdoor Environments.* Berkeley, Calif.: MIG Communications.

Moore, R.C., S.M. Goltsman, and D.S. Iacofano (1992). *Play for All Guidelines: Planning, Design, and Management of Outdoor Play Settings for All Children.* Berkeley, Calif.: MIG Communications.

Morrison, M.L., and M.E. Fise (1992). *Report and Model Law on Public Play Equipment and Areas.* Washington, D.C.: Consumer Federation of America.

National Recreation and Park Association (1992). *Playground Equipment for Public Use Continuum of Skills and Size Differences of Children Age Two to Twelve.* Arlington, Va.: NRPA.

PLAE, Inc. (1993). *Universal Access to Outdoor Recreation: A Design Guide.* Berkeley, Calif.: MIG Communications.

Ramsey, L.F., and J.D. Preston (1990). *Impact Attenuation Performance of Playground Surfacing Materials.* Washington, D.C.: U.S. Consumer Product Safety Commission.

Ratte, D.J., M.L. Morrison, and N.D. Lerner (1990). *Development of Human Factors Criteria for Playground Equipment Safety; COMSIS Corporation.* Washington, D.C.: U.S. Consumer Product Safety Commission.

Riley, B. (1994). *Getting Pesticides Out of Our Schools.* Eugene, Ore.: Northwest Coalition for Alternatives to Pesticides.

Tinsworth, D.K., and J.T. Kramer (1990). *Playground Equipment–Related Injuries and Deaths.* Washington, D.C.: U.S. Consumer Product Safety Commission.

U.S. Consumer Product Safety Commission (1994). *Handbook for Public Playground Safety.* Washington, D.C.: CPSC.

Wallach, F., ed. (1992). *State Regulations Focused on Playgrounds and Supervision.* Arlington, Va.: National Recreation and Park Association.